Advances in Intelligent and Soft Computing 86

Editor-in-Chief: J. Kacprzyk

Advances in Intelligent and Soft Computing

Editor-in-Chief

Prof. Janusz Kacprzyk
Systems Research Institute
Polish Academy of Sciences
ul. Newelska 6
01-447 Warsaw
Poland
E-mail: kacprzyk@ibspan.waw.pl

Further volumes of this series can be found on our homepage: springer.com

Vol. 73. E. Corchado, P. Novais, C. Analide, J. Sedano (Eds.)
Soft Computing Models in Industrial and Environmental Applications, 5th International Workshop (SOCO 2010), 2010
ISBN 978-3-642-13160-8

Vol. 74. M.P. Rocha, F.F. Riverola, H. Shatkay, J.M. Corchado (Eds.)
Advances in Bioinformatics, 2010
ISBN 978-3-642-13213-1

Vol. 75. X.Z. Gao, A. Gaspar-Cunha, M. Köppen, G. Schaefer, and J. Wang (Eds.)
Soft Computing in Industrial Applications, 2010
ISBN 978-3-642-11281-2

Vol. 76. T. Bastiaens, U. Baumöl, and B.J. Krämer (Eds.)
On Collective Intelligence, 2010
ISBN 978-3-642-14480-6

Vol. 77. C. Borgelt, G. González-Rodríguez, W. Trutschnig, M.A. Lubiano, M.Á. Gil, P. Grzegorzewski, and O. Hryniewicz (Eds.)
Combining Soft Computing and Statistical Methods in Data Analysis, 2010
ISBN 978-3-642-14745-6

Vol. 78. B.-Y. Cao, G.-J. Wang, S.-Z. Guo, and S.-L. Chen (Eds.)
Fuzzy Information and Engineering 2010
ISBN 978-3-642-14879-8

Vol. 79. A.P. de Leon F. de Carvalho, S. Rodríguez-González, J.F. De Paz Santana, and J.M. Corchado Rodríguez (Eds.)
Distributed Computing and Artificial Intelligence, 2010
ISBN 978-3-642-14882-8

Vol. 80. N.T. Nguyen, A. Zgrzywa, and A. Czyzewski (Eds.)
Advances in Multimedia and Network Information System Technologies, 2010
ISBN 978-3-642-14988-7

Vol. 81. J. Düh, H. Hufnagl, E. Juritsch, R. Pfliegl, H.-K. Schimany, and Hans Schönegger (Eds.)
Data and Mobility, 2010
ISBN 978-3-642-15502-4

Vol. 82. B.-Y. Cao, G.-J. Wang, S.-L. Chen, and S.-Z. Guo (Eds.)
Quantitative Logic and Soft Computing 2010
ISBN 978-3-642-15659-5

Vol. 83. J. Angeles, B. Boulet, J.J. Clark, J. Kovecses, and K. Siddiqi (Eds.)
Brain, Body and Machine, 2010
ISBN 978-3-642-16258-9

Vol. 84. Ryszard S. Choraś (Ed.)
Image Processing and Communications Challenges 2
ISBN 978-3-642-16294-7

Vol. 85. Á. Herrero, E. Corchado, C. Redondo, and Á. Alonso (Eds.)
Computational Intelligence in Security for Information Systems 2010
ISBN 978-3-642-16625-9

Vol. 86. E. Mugellini, P.S. Szczepaniak, M.C. Pettenati, and M. Sokhn (Eds.)
Advances in Intelligent Web Mastering – 3
ISBN 978-3-642-18028-6

Elena Mugellini, Piotr S. Szczepaniak,
Maria Chiara Pettenati, and Maria Sokhn (Eds.)

Advances in Intelligent Web Mastering – 3

Proceedings of the 7th Atlantic Web Intelligence Conference, AWIC 2011, Fribourg, Switzerland, January, 2011

Springer

Editors

Elena Mugellini
University of Applied Sciences
of Western Switzerland
Fribourg, Bd de Pérolles 80
1705 Fribourg
Switzerland
E-mail: elena.mugellini@hefr.ch

Maria Chiara Pettenati
University of Florence
Via S. Marta, 3
50139 Firenze
Italy
E-mail:
pettenati.mariachiara@gmail.com

Prof. Dr. Piotr S. Szczepaniak
Technical University of Lodz
ul. Wólczanska 215
90-924 Lodz
Poland
E-mail: piotr@ics.p.lodz.pl

Maria Sokhn
University of Applied Sciences
of Western Switzerland
Bd de Pérolles 80
1705 Fribourg
Switzerland
E-mail: maria.sokhn@hefr.ch

ISBN 978-3-642-18028-6 e-ISBN 978-3-642-18029-3

DOI 10.1007/978-3-642-18029-3

Advances in Intelligent and Soft Computing ISSN 1867-5662

Typeset & Cover Design: Scientific Publishing Services Pvt. Ltd., Chennai, India.

Printed on acid-free paper
5 4 3 2 1 0
springer.com

Preface

The Atlantic Web Intelligence Conference brings together scientists, engineers, computer users, and students to exchange and share their experiences, new ideas, and research results about all aspects (theory, applications and tools) of intelligent methods applied to Web based systems, and to discuss the practical challenges encountered and the solutions adopted. Previous AWIC events were held in Spain – 2003, Mexico – 2004, Poland – 2005, Israel – 2006, France – 2007 and Czech Rep. – 2009.

This year, the 7th Atlantic Web Intelligence Conference (AWIC'2011) is held during January 26-28, 2011, at the University of Applied Sciences of Fribourg, Switzerland. AWIC2011 is organized by the Multimedia Information System Group (MISG), Institute of the Technologies of Information and Communication (iTIC) of the University of Applied Sciences of Fribourg.

The conference has attracted submissions from several parts of the world and each paper was reviewed by two or three reviewers. The diversity of the topics dealt within submitted papers and the quality of the accepted ones in addition to the keynote speakers (*Philippe Cudré-Mauroux,* University of Fribourg, department of informatics, Fribourg, Switzerland, *Sławomir Zadrożny*, Systems Research Institute, Polish Academy of Sciences, Poland and *Vicenzo Loia*, Univesrsity of Salerno, Italy) make this conference an interesting and rich event.

We would like to express our sincere gratitude to the program committee, local organizing committee and all the referees who helped us evaluating the papers and making AWIC2011 a very successful scientific event. Grateful appreciation is expressed to Professor Janusz Kacprzyk, Editor of the series publishing this book as well as to the Springer team for its excellent work.

We hope that every reader will find interest and inspiration along this book.

Fribourg, January 2011 The editors

Contents

Part I
Invited Lectures

Fuzzy Ontologies and Fuzzy Markup Language: A Novel Vision in Web Intelligence

Vincenzo Loia

Abstract. Semantic Web and Ontologies has increased the interest in the use of Knowledge-based systems in order to allow automated processing of, and reasoning with, information on the Web. However, it is widely pointed out that standard ontologies are not sufficient to deal with imprecise and vague knowledge for some real world applications, but the fuzzy ontologies can effectively model data and knowledge with uncertainty. In particular, in this paper will be introduced a collection of real-world applications based on the integration of different web-oriented frameworks such as the *ontology-based intelligent fuzzy agents (OIFAs)* and the *Fuzzy Markup Language (FML)* capable of generating fuzzy inference mechanisms and semantic decision making systems for an efficient modeling of real scenarios. In detail, hereafter our web intelligence approach will be applied to medical semantic decision making systems, computer go framework and so on.The experimental results show that the proposed method is feasible for different real-word scenarios.

1 Introduction

Intelligent Agents concepts and Web technologies enable novel and suitable knowledge representation approaches able to develop and design automated intelligent processing capabilities almost comparable with traditional human subjective evaluation and reasoning. These innovative knowledge representation approaches strongly exploit advances in semantic web and Internet technologies that have accelerated the growth of the research on the intelligent multi-agent framework by proving that it is a must to combine different ontologies among them in order to make an agent with semantic related components. As well, it is widely pointed out that classical ontologies are not sufficient to represent the real-world both vague and imprecise

Vincenzo Loia
Department of Mathematics and Computer Science,
University of Salerno, Italy
e-mail: loia@unisa.it

E. Mugellini et al. (Eds.): Advances in Intelligent Web Mastering - 3, AISC 86, pp. 3–10.
springerlink.com

information because the information usually involves vagueness or imprecision in many real-world applications such as medical diagnosis, and so on. Therefore, fuzzy ontologies are strongly suitable to handle real-world knowledge and, as consequence, to design advanced intelligent agent systems for different application domains such as medical decision making systems, computer go frameworks, Capability Maturity Model Integration assessment and so on. In this context, the Fuzzy Markup Language (FML) represents one of the most important result because it allows fuzzy scientists to express their ideas in abstract and interoperable way by improving their productivity and, at the same time, increasing the average quality of their works. In particular, FML programs code fuzzy controllers [1][2][3] that are control system based on fuzzy logic, a mathematical framework that transform input knowledge through logical variables that take on continuous values belonging to the so-called *fuzzy sets*[4]. The XML nature of FML enable a direct analysis of fuzzy ontologies in FML programs as will be shown in next sections.

2 Fuzzy Ontologies and Fuzzy Markup Language

This section is devoted to introduce Fuzzy Ontologies and FML. The joint exploitation of these technologies will allow web intelligence designers to develop fuzzy inference mechanisms and semantic decision making systems for an efficient modeling of real scenarios.

2.1 *Fuzzy Ontologies*

In this section, we present a fuzzy ontology definition[5]:

Definition 1. A *Fuzzy Ontology* is defined as the quintuple $OF = \{C,I,R,F,A\}$ where:

- I is the set of individuals, also called instances ofthe concepts;
- C is the set of concepts. Each concept $c \in C$ is a fuzzy set on the domain of instances $c : I \rightarrow [0,1]$. The set of entities of the fuzzy ontology will be indicated by E, i.e., $E = C \cup I$;
- R is the set of relations. Each $r \in R$ is a n-ary fuzzy relation on the domain of entities $r : E^n \rightarrow [0,1]$. A special role is held by the taxonomic relation $T : E^2 \rightarrow [0,1]$ which identifies the fuzzy subsumption relation among the entities of the ontology;
- F is the set of the fuzzy relations on the set of entities E and a specific domain contained in $D = \{integer, string, \ldots\}$. In detail, they are n-ary functions such that each element $f \in F$ is a relation $f : E^{(n-1)} \times P \rightarrow [0,1]$ where $P \in D$;
- A is the set of axioms expressed in an proper logical language, i.e., predicates that constrain the meaning of concepts, individuals, relationships and functions.

Fuzzy ontologies have been applied in different application domains. Figure 1 shows a fuzzy ontology capable of representing the knowledge related to a personal diet plan.

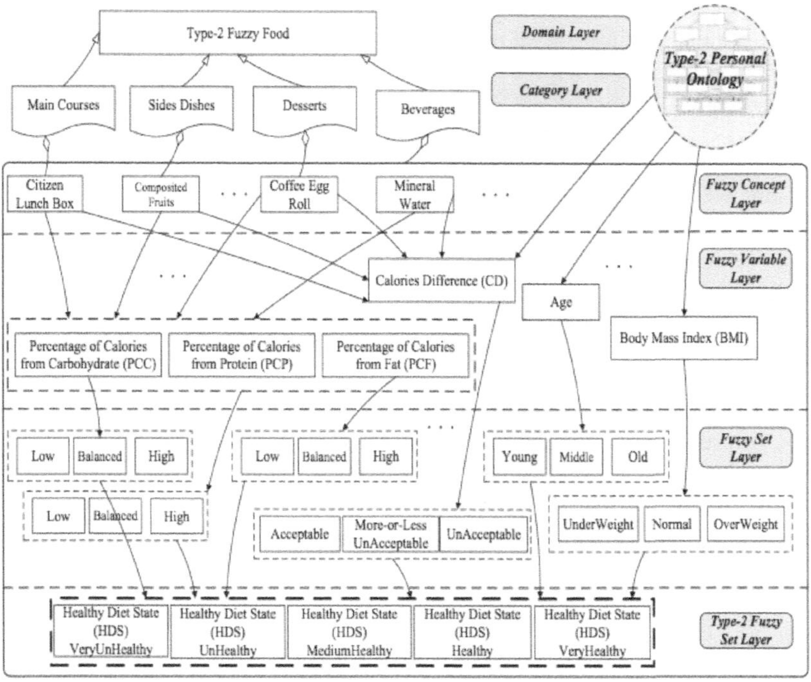

Fig. 1 A fuzzy ontology for diet assessment

In order to design advanced decision making systems, it is necessary to embed fuzzy knowledge in a high-level inference engine able to analyze fuzzy instances and infer novel information by exploring fuzzy relationships through well-defined computational intelligence methods such as the *fuzzy control*. Fuzzy Markup Language (FML) is an advanced method based on web technologies used to model fuzzy controllers in a transparent and hardware independent way and, for these reasons, it is suitable to analyze fuzzy information contained in fuzzy ontologies. Hereafter, FML will be introduced together with a collection of real-world applications in order to prove the suitability of the joint exploitation of fuzzy controllers and fuzzy ontologies.

2.2 *Fuzzy Markup Language*

Fuzzy Logic Controller (FLC) [6][7] lets controller designers to describe complex systems using their knowledge and experience by means of linguistic rules [8]. It does notrequire system modeling or complex math equations governing the relationship between inputs and outputs as it happens for other controller design methodologies. FLC typically takes a few rules to describe systems that may require several of lines of conventional software. However, in spite of these unquestionable advantages, the real design of FLCs strongly depends upon the methodologies modeling the application domain knowledge. In this paper, an alternative vision of FLCs implementation, allowing the designer to model the controllers in an independent methodology way, is presented [9]. XML is the most widely used tool for the heterogeneous data modeling, and for this reason, the labeled tree[10] can be viewed as the contact point between FLC and its XML representation. The merging among FLC, labeled tree theory, and XML methodology is called Transparent Fuzzy Control. From a bottom-up point of view, a FLC can be view as a collection of fuzzy concepts and fuzzy rules composing the fuzzy knowledge base and fuzzy rule base, respectively. The XML representation of FLC allows human to model the controller in a human-readable and abstract way. The XML-based language modeling FLCs is named Fuzzy Markup Language (FML) [11]. FML has been designed and implemented by Giovanni Acampora in his PhD thesis under the supervision of Prof. Vincenzo Loia. An FML implementation is based on the employment of tags to model different parts of fuzzy controller. The root of the structure of FML models the controller node and the FML tag <FUZZYCONTROL> is used. The FML sub-trees use the FML tag <KNOWLEDGEBASE> to model the Knowledge Base Component. On the other hand, the FML right sub-trees define the Rule Base Component and the FML tag <RULEBASE> is used. An example of FML program is shown in figure 2.

3 Fuzzy Ontologies and FML: Real-World Applications

In this section, three real-world applications, where the effectiveness of exploiting together fuzzy control and fuzzy ontology is shown, will be described. The three selected applications belong to two particular domains: the medical environment (the diabetes and diet applications) and the computer games (go board game).

Starting from consideration that the classical ontologies are not able to model vague and imprecise information, in [12] the authors employ fuzzy ontologies and logic fuzzy principles represented by FML in a transparent way in order to build up an ontology-based intelligent fuzzy agent (OIFA) to apply to diabetes semantic decision making domain. Diabetes is a disease in which your body cannot properly use the food you eat for energy. The goal of treating diabetes is to keep the glucose level as near to normal as possible. The aim of the designed OIFA is to realize the glucose level control following a method which is composed of a FML generating mechanism, a FML parser, a fuzzy inference mechanism, and a semantic decision making mechanism. First, the FML generating mechanism is responsible to represent a proposed fuzzy diabetes ontology (FDO) and then storing the generated FML-based definitions to the diabetes knowledge base and rule base. Next, based on the retrieved FML-based definitions, the FML parser analyzes the definitions and translates them into the desired knowledge file format. The fuzzy inference mechanism then executes the fuzzy inference to infer the possibility of suffering from diabetes according to the output knowledge files of the FML parser. Finally, the semantic decision making mechanism translates the inferred results into

```
<?xml version="1.0" encoding="UTF-8" standalone="no"?>
<FuzzyController ip="140.133.33.2" name="Healthy Diet">
    <KnowledgeBase>
        <FuzzyVariable domainleft="0.0" domainright="100.0" name="PCC" scale="%" type="input">
            <FuzzyTerm complement="false" name="Low">
                <TrapezoidShape Param1="0.0" Param2="0.0" Param3="50.0" Param4="55.0"/>
            </FuzzyTerm>
            <FuzzyTerm complement="false" name="Balanced">
                <TrapezoidShape Param1="50.0" Param2="55.0" Param3="65.0" Param4="70.0"/>
            </FuzzyTerm>
            <FuzzyTerm complement="false" name="High">
                <TrapezoidShape Param1="65.0" Param2="70.0" Param3="100.0" Param4="100.0"/>
            </FuzzyTerm>
        </FuzzyVariable>
        <FuzzyVariable domainleft="0.0" domainright="250.0" name="CD" scale="%" type="input">
            <FuzzyTerm complement="false" name="Acceptable">
                <TrapezoidShape Param1="0.0" Param2="0.0" Param3="50.0" Param4="100.0"/>
            </FuzzyTerm>
            <FuzzyTerm complement="false" name="MoreOrLessAcceptable">
                <TrapezoidShape Param1="70.0" Param2="100.0" Param3="150.0" Param4="200.0"/>
            </FuzzyTerm>
            <FuzzyTerm complement="false" name="UnAcceptable">
                <TrapezoidShape Param1="150.0" Param2="200.0" Param3="250.0" Param4="250.0"/>
            </FuzzyTerm>
        </FuzzyVariable>
...
        <FuzzyVariable accumulation="MAX" defaultValue ="0.0" defuzzifier="COG" domainleft="0.0" domainright="1.0"
name="HDS" scale="" type="output">
            <FuzzyTerm complement="false" name="VeryUnHealty">
                <TrapezoidShape Param1="0.0" Param2="0.0" Param3="0.1" Param4="0.25"/>
            </FuzzyTerm>
            <FuzzyTerm complement="false" name="UnHealthy">
                <TrapezoidShape Param1="0.1" Param2="0.25" Param3="0.25" Param4="0.5"/>
            </FuzzyTerm>
            <FuzzyTerm complement="false" name="MediumHealthy">
                <TrapezoidShape Param1="0.25" Param2="0.5" Param3="0.5" Param4="0.75"/>
            </FuzzyTerm>
            <FuzzyTerm complement="false" name="Healthy">
                <TrapezoidShape Param1="0.5" Param2="0.75" Param3="0.75" Param4="0.9"/>
            </FuzzyTerm>
            <FuzzyTerm complement="false" name="VeryHealthy">
                <TrapezoidShape Param1="0.75" Param2="0.9" Param3="1.0" Param4="1.0"/>
            </FuzzyTerm>
        </FuzzyVariable>
    </KnowledgeBase>
    <RuleBase activationMethod="MIN" andMethod="MIN" name="RuleBase1" orMethod="MAX" type="mamdani">
        <Rule connector="and" name="RULE1" operator="MIN" weight="1.0">
            <Antecedent>
                <Clause>
                    <Variable> PCC</Variable>
                    <Term>Low</Term>
                </Clause>
                <Clause>
                <Clause>
                    <Variable> CD</Variable>
                    <Term>Acceptable</Term>
                </Clause>
            </Antecedent>
...
            <Consequent>
                <Clause>
                    <Variable> HDS</Variable>
                    <Term>UnHealthy</Term>
                </Clause>
            </Consequent>
...
</RuleBase>
</FuzzyController>
```

Fig. 2 FML code for diet system

the semantic descriptions and stores them in the diabetes decision making repository. The planned ontology-based intelligent fuzzy agent was implemented using the C++ Builder 2007 Programming language. The performance of the proposed agent is evaluated according to the criteria such as accuracy, precision, and recall. The experimental results show that the proposed method is feasible for diabetes semantic decision-making since it is able to analyze data and further translate them into knowledge to simulate the medical humans thinking process.

The second selected application concerns the diet assessment [13]. Starting from consideration that a system capable of providing personalized healthy diets could help an individual to better live his life, the work exploits fuzzy ontology concepts in order to determine individual dietary status and realize an intelligent agent able to analyze this status and propose the most suitable and healthy dietary pattern. Indeed, in this context, a computer-assisted can provide recommendations based on an individuals usual eating habits, food preferences, and state of change and allow health practitioners to focus their time on the nutritional needs of the individual rather than the coding and sorting of the dietary data. In short, the proposed approach consists in a novel type-2 fuzzy ontology, including a type-2 fuzzy food ontology and a type-2 fuzzy markup language (FML)-based ontology, and in the designing of a FML2-based diet assessment agent. In particular, in this work, together with fuzzy ontologies, the type-2 version of FML is used to model a type-2 fuzzy control, i.e, a controller which exploits type-2 fuzzy sets [14]. More in details, the FML2-based diet assessment agent includes a type-2 knowledge engine, a type-2 fuzzy inference engine, a diet assessment engine, and a semantic analysis engine. In the proposed method, first, the nutrition facts of various kinds of food are collected from the Internet and the convenience stores. Next, the domain experts construct the type-2 fuzzy ontology, and then the involved subjects are requested to input the different food eaten. Finally, the proposed FML2-based diet assessment agent displays the diet assessment of the food eaten based on the constructed type-2 fuzzy ontology. Using the generated semantic analysis, people can obtain health information about what they eat, which can lead to a healthy lifestyle and healthy diet. Experimental results show that the proposed approach works effectively where the proposed system can provide a diet health status, which can act as a reference to promote healthy living.

Finally, by leaving medical environment, in [15], the authors propose an FML-based type-2 fuzzy ontology for computer Go knowledge representation. The game of Go is one of the last board games where the strongest humans are still able to easily win against computers in 19x19 games. The work is based on the idea that if the computer can learn more knowledge and strategy from professional Go players through a constructed ontology, then the computer Go will approach the level of professional go player very fast. For this reason, the proposed approach consists in building a type-2 fuzzy ontology (T2FO) based on type-2 fuzzy sets (T2FSs) to describe the fuzzy concepts and fuzzy relations in computer go game domain. Also in this work as the previous one, the type-2 FML version is exploited. The proposed T2FO stores T2FSs and is an extended version of the fuzzy ontology which contains six layers, including a domain layer, a category layer, a fuzzy concept layer, a fuzzy variable layer, a type-l fuzzy set layer, and a type-2 fuzzy set (T2FS) layer. Once the fuzzy ontology experts construct the fuzzy ontology through the type-2 fuzzy set construction mechanism, a type-2 fuzzy set inference mechanism infers the winning rate of the game by basing on the fuzzy ontology repository. Based on the FML, an FML editor is used to construct the important knowledge base and rule base of the type-2 fuzzy set inference mechanism. Future experiments will be executed in order to demonstrate the effectiveness of the proposed approach to increase the winning rate of the game.

4 Conclusion and Future Works

This paper introduces a novel vision in Web Intelligence. We showed how the joint exploitation of fuzzy ontologies and Fuzzy Markup Language enables a simple and direct approach towards the intelligent design of knowledge based frameworks and advanced decision making systems. In the future the same approach will be integrated with other computational intelligence techniques such as neural networks and evolutionary algorithms in order to propose innovative and adaptive knowledge based systems.

References

1. Takagi, T., Sugeno, M.: Fuzzy identification of systems and its applications to modeling and control. IEEE Transactions Systems, Man & Cybernetics 15(1), 116–132 (1985)
2. Mamdani, E.H.: Applications of fuzzy algorithms for simple dynamic plants. Proc. IEE 121, 1585–1588 (1974)
3. Acampora, G., Loia, V., Vitiello, A.: Enhancing Transparent Fuzzy Controllers through Temporal Concepts: An Application to Computer Games. In: International Workshop on Computer Games (IWCG 2010), Conference on Technologies and Applications of Artificial Intelligence (TAAI 2010) (November 2010)
4. Zadeh, L.A.: Fuzzy sets. Information and Control 8, 338–353 (1965)
5. Calegari, S., Ciucci, D.E.: Fuzzy Ontology, Fuzzy Description Logics and Fuzzy-OWL. In: Masulli, F., Mitra, S., Pasi, G. (eds.) WILF 2007. LNCS (LNAI), vol. 4578, pp. 118–126. Springer, Heidelberg (2007)
6. Lee, C.C.: Fuzzy Logic in Control System: Fuzzy Logic Controller - Part I and Part II. IEEE Transactions on SMC 20, 404–435 (1990)
7. Acampora, G., Loia, V., Vitiello, A.: Hybridizing fuzzy control and timed automata for modeling variable structure fuzzy systems. In: IEEE International Conference on Fuzzy Systems (FUZZ 2010), July 18-23, pp. 1–8 (2010), doi:10.1109/FUZZY.2010.5584393.
8. Acampora, G., Lee, C.S., Wang, M.H.: FML-Based Ontological Agent for Healthcare Application with Diabetes. In: Web Intelligence/IAT Workshops, pp. 413–416 (2009)
9. Acampora, G., Loia, V.: An Open Integrated Environment for Transparent Fuzzy Agents Design. In: IFIP International Federation for Information Processing, vol. 275, pp. 249–255. Springer, Boston (2008)
10. Wang, Y.L., Chen, H.C., Liu, W.K.: A Parallel Algorithm for Constructing a Labeled Tree. IEEE Transactions on Parallel and Distributed Systems 8, 1236–1240

11. Acampora, G., Loia, V.: Fuzzy control interoperability and scalability for adaptive domotic framework. IEEE Transactions on Industrial Informatics 1(2), 97–111 (2005)
12. Lee, C.S., Wang, M.H., Acampora, G., Loia, V., Hsu, C.Y.: Ontology-based intelligent fuzzy agent for diabetes application. In: IEEE Symposium on Intelligent Agents (IA 2009), pp. 16–22 (2009), doi:10.1109/IA.2009.4927495.
13. Lee, C.S., Wang, M.H., Acampora, G., Loia, V., Hsu, Hagras, H.: Diet assessment based on type-2 fuzzy ontology and fuzzy markupl language. Int. J. Intell. Syst. 25(12), 1187–1216 (2010), http://dx.doi.org/10.1002/int.v25:12, doi:10.1002/int.v25:12
14. Hagras, H.: Type -2 FLCs: A new Generation of Fuzzy Controllers. IEEE Computational Intelligence Magazine 2, 30–43 (2007)
15. Lee, C.S., Wang, M.H., Yan, Z.R., Chen, Y.J., Doghmen, H., Teytaud, O.: FML-based type-2 fuzzy ontology for computer go knowledge representation. In: 2010 International Conference on System Science and Engineering (ICSSE), July 1-3, pp. 63–68 (2010), doi:10.1109/ICSSE.2010.5551703

Loose Ontological Coupling and the Social Semantic Web

Philippe Cudré-Mauroux

Abstract. Best practices for the publication of Semantic Web data currently place an unacceptably high burden on the end-user, who is supposed to locate and embrace third-party ontological structures prior to publishing any information. In this paper, I argue for a different publication paradigm where end-users are encouraged to publish potentially incomplete or conflicting information according to their own local context, and where heterogeneous data is consolidated *a posteriori* through bottom-up, decentralized processes. This approach simplifies both the publication and curation processes, while opening the door to pay-as-you-go knowledge integration and human-centered social semantics. However, it also profoundly alters the semantics of the overall resulting system.

Keywords: Social Semantics, Human-Centered Computing, Loosely-Coupled Semantics, Web of Data, Emergent Semantics.

1 Introduction

In the last few years, collaborative tools and social computing have profoundly altered the nature of the World Wide Web. A significant fraction of the online population is today used to contributing content on a wide variety of platforms, ranging from video sharing websites to online encyclopedia or tagging portals. End-users collaborate on those platforms in loosely-coupled ways: they are able to contribute content independently of what previous users have uploaded, and can often comment on or even revise the information generated by other users without requiring their direct consent. The

Philippe Cudré-Mauroux
eXascale Infolab,
University of Fribourg – Switzerland
`http://diuf.unifr.ch/xi`

E. Mugellini et al. (Eds.): Advances in Intelligent Web Mastering - 3, AISC 86, pp. 11–15.
springerlink.com

systematic resort to loosely-coupled interactions between users and content on one hand and between various pieces of uploaded content on the other hand is one of the key principles fostering the wide adoption of such collaborative platforms, as it considerably lowers the prerequisites for publishing information on the social Web.

Publishing and reusing content on the Semantic Web is today much harder than on the social Web. Beyond the lack of user-friendly tools or portals, one of the main requirements hampering the wide adoption of Semantic Web languages is the implicit prerequisite dictating the reuse of preexisting ontologies. Ontologies are often defined through complex formal constructs (such as those provided by OWL [7]) and serialized in indecipherable machine-readable formats like XML. Expecting end-users to i) locate suitable ontologies on the Web and ii) embrace (part of) those complex ontologies for their own needs is thus not realistic in general. As a matter of fact, upper or standardized ontologies have been debated for years but have had very little impact so far—limited to well-organized communities (e.g., genomics ontologies) or simplistic domains (e.g., FOAF [1]).

Such best-practices promoting the adoption of centralized conceptualizations and systematic knowledge-reuse were promulgated to avoid all inconsistencies through *a priori* consensus, and to foster interoperability through top-down, mediated integration techniques. In the following, I argue for a different Web ecology, where end-users are encouraged to define their own schemas or ontologies, independently of what other users or communities might have already suggested, and where semantics are consolidated *a posteriori* through local, probabilistic, and bottom-up interactions.

2 Loosely-Coupled Ontologies

Tightly-coupled semantics revolving around standard schemas and ontologies are often considered as a necessity to support inference capabilities and enable large-scale knowledge-reuse. Promoting a multiplicity of loosely-coupled, disparate and potentially conflicting user-defined ontologies has a number of advantages, however, including:

Low Publishing and Maintenance Costs: Publishing Semantic Web information in isolation without resorting to third-party concepts considerably lowers the publishing costs. End-users can in that case control the publication of their data end-to-end and can axiomatize previously-defined database schemas or taxonomies in their own ways to locally create the ontological structures corresponding to their own context. The maintenance process is significantly simplified as well, since the user is the sole owner of the ontologies it uses and can make them evolve as he sees fit.

Pay-As-You-Go Integration: Adopting an existing ontology to publish heterogeneous or pre-existing data often requires an important upfront effort to integrate schemas or ontologies *a priori* and massage the various data sets until they all conform to the unique standardized conceptualization. Loosely-coupled ontologies, on the other hand, promote decentralized, *Peer Data Management* [4] integration techniques where information is integrated dynamically through pairwise mappings, without resorting to any central component.

Personalized Semantics: Finally and perhaps most importantly, loosely-coupled semantics open the door to personalized and contextualized semantics on the Web. Standard ontologies *de facto* restrict the ways users can model or express knowledge. Decentralized semantics, on the other hand, would foster human-centered, community-based *social semantics* where individuals can express their preferences through various ontological commitments (e.g., "those two concepts should not be related from my perspective" or "this painting is a post-impressionist piece of art") and can be related to other users through some shared or semantically related conceptualizations, but also through their ontological specificities and semantic conflicts.

3 Emergent Semantics

Current Semantic Web formalisms were not conceived to be integrated *a posteriori*. They presuppose a global and strictly coherent ontological commitment through which all sources agree on the local semantics of their data. Thus, if conflicts arise and are used to foster personalized semantics, how would the semantics of the overall system (i.e., the Semantic Web) be modeled? Besides classical, top-down semantics, various paradigms exist to define semantics in evolutionary manners (e.g., [10]). Emergent semantics [2], in particular, seem to be particularly well-suited to the dynamic and decentralized nature of the Web.

The term *Emergent Semantics* refers to a set of principles and techniques analyzing the evolution of decentralized semantic structures in large-scale distributed information systems. Emergent semantics approaches model the semantics of a distributed system as an ensemble of relationships between syntactic structures. They consider both the representation of semantics and the discovery of the proper interpretation of symbols as the result of a self-organizing process performed by distributed agents exchanging symbols and having utilities dependent on the proper interpretation of the symbols [3]. This is a complex systems perspective on the problem of dealing with semantics.

Emergent semantics expresses semantics through purely syntactic, recursive domains. The notions underlying emergent semantics are rooted in computational linguistics works relating semantics to the analysis of syntactic constructs [8, 9]. In a large scale distributed environment such as the Semantic Web, the aim would be to have local ontological structures interoperate irrespective of their initial local context. To that aim, human or computational agents would have to map local concepts (carrying the meaning as initially defined in its *base* schema or ontology) to the concepts of other ontologies with which it wants to interoperate. Hence, a relationship between local and distant symbols is established. This relationship may be considered as another form of semantics, independent of the initial semantics of the symbols.

Assuming that local ontologies are progressively mapped to distant ontological elements through various relations (e.g., formal *Same As* predicates, probabilistic mappings or loosely-defined *Like* links), the original *human assigned* semantics would loose its relevance; from an agent's perspective, *new* semantics would then result from the relationships linking the local constructs to their environment. This is a novel way of providing semantics to local ontologies relative to the symbols of other ontologies with which they interact. Typically, this type of semantic representation is distributed such that no single source holds a complete representation of a generally agreed-upon semantics. The bottom-up reconciliation or augmentation processes required in this context can be applied at the ontological or schematic level [5] as well as on instance-level data [6]. Recent efforts towards fuzzy [11] or incomplete [12] reasoning paradigms would also prove to be essential in that sense, in order to complement probabilistic and Bayesian analyses.

4 Conclusions

Current practices for the publication of Semantic Web information place an unacceptably high burden on the end-user, who is supposed to locate and univocally embrace third-party ontological structures prior to publishing its own information. Loosely-coupled, personalized ontologies would enable simplified publication and curation processes, while opening the door to pay-as-you-go knowledge integration and human-centered social semantics. In that context, missing or conflicting information would prove to be essential in order to provide contextualized or personalized semantics and to relate local symbols to distant ontological elements. The resulting semantics of the overall Semantic Web would then indubitably change, but could be captured by novel paradigms such evolutionary semiotics or emergent semantic principles.

References

1. Brickley, D., Miller, L.: FOAF Vocabulary Specification 0.91, `http://xmlns.com/foaf/spec/`
2. Cudré-Mauroux, P.: Emergent Semantics. EPFL & CRC Press (2008)
3. Cudré-Mauroux, P.: Emergent Semantics. In: Özsu, T.M., Liu, L. (eds.) Encyclopedia of Database Systems. Springer, Heidelberg (2009)
4. Cudré-Mauroux, P.: Peer Data Management System. In: Özsu, T.M., Liu, L. (eds.) Encyclopedia of Database Systems. Springer, Heidelberg (2009)

5. Cudré-Mauroux, P., Aberer, K., Feher, A.: Probabilistic Message Passing in Peer Data Management Systems. In: International Conference on Data Engineering, ICDE (2006)
6. Cudré-Mauroux, P., Haghani, P., Jost, M., Aberer, K., de Meer, H.: idMesh: Graph-Based Disambiguation of Linked Data. In: International World Wide Web Conference, WWW (2009)
7. McGuinness, D., van Harmelen, F. (ed.): OWL Web Ontology Language Overview. W3C Recommendation (February 2004), `http://www.w3.org/TR/owl-features/`
8. Rapaport, W.: Syntactic Semantics: Foundations of Computational Natural-Language Understanding. In: Aspects of Artificial Intelligence, pp. 81–131. Kluwer Academic Publishers, Dordrecht (1988)
9. Rapaport, W.: What Did You Mean by That? Misunderstanding, Negotiation, and Syntactic Semantics. Minds and Machines 13(3), 397–427 (2003)
10. Steels, L.: Semiotic dynamics for embodied agents. IEEE Intelligent Systems 21(3), 32–38 (2006)
11. Stoilos, G., Stamou, G., Pan, J.: Fuzzy extensions of owl: Logical properties and reduction to fuzzy description logics. International Journal of Approximate Reasoning 51(6), 656–679 (2010)
12. Stoilos, G., Grau, B.C., Horrocks, I.: Completeness guarantees for incomplete reasoners. In: International Semantic Web Conference, ISWC (2010)

Part II
Regular Papers

Further Experiments in Sentiment Analysis of French Movie Reviews

Hatem Ghorbel and David Jacot

Abstract. In sentiment analysis of reviews we focus on classifying the polarity (positive, negative) of conveyed opinions from the perspective of textual evidence. Most of the work in the field has been intensively applied on the English language and only few experiments have explored other languages. In this paper, we present a supervised classification of French movie reviews where sentiment analysis is based on some shallow linguistic features such as POS tagging, chunking and simple negation forms. In order to improve classification, we extracted word semantic orientation from the lexical resource SentiWordNet. Since SentiWordNet is an English resource, we apply a word-translation from French to English before polarity extraction. Our approach is evaluated on French movie reviews, obtained results showed that shallow linguistic features has significantly improved the classification performance with respect to the bag of words baseline.

Keywords: Sentiment analysis, Opinion Mining, Polarity classification, Supervised learning, Linguistic features.

1 Introduction

Sentiment analysis is an emerging discipline whose goal is to analyze textual content from the perspective of the opinions and viewpoints they hold. A large number of studies have focused on the task of defining the polarity of a document which is by far considered as a classification problem: decide to which class a document is attributed; class of positive or negative polarity [HR10, WK09, PLV02].

Hatem Ghorbel · David Jacot
University of Applied Sciences Western Switzerland,
Haute Ecole Arc Ingénierie, St-Imier,
Switzerland
e-mail: hatem.ghorbel@he-arc.ch,
david.jacot@master.hes-so.ch

E. Mugellini et al. (Eds.): Advances in Intelligent Web Mastering - 3, AISC 86, pp. 19–28.
springerlink.com

Most of the work in the field has been intensively applied on the English language [Tur02, WK10]. For this purpose, English corpora and resources (such as MPQL [WWH05], Movie Review Data [PLV02], SentiWordNet [ES06] and WordNet-Affect [SV04]) have been constructed to aid in the process of automatic supervised and unsupervised polarity classification of textual data. Nevertheless, still very few experiments are applied on other languages.

In this context, we address in this paper the issue of polarity classification applied on French movie reviews. We used a supervised learning approach where we trained the classifier on annotated data of French movie reviews extracted from the web. As classification features, beyond the word unigrams feature taken as the baseline in our experiments, we extracted further linguistic features including lemmatized unigrams, POS tags, simple negation forms and semantic orientation of selected POS tags. The latter is extracted from the English lexical resource SentiWordNet after applying a word-translation from the French to English.

The main goal of our experiments is firstly to confirm that the incorporation of shallow linguistic features into the polarity classification task could significantly improve the results. Secondly, to address the problem of loss of precision in defining the semantic orientation of word unigrams from English lexical resources, mainly due to the intermediate process of word-translation from French to English correlated with further issues such as sense disambiguation.

In the rest of the paper, we first shortly describe the previous work in the field of sentiment analysis and polarity classification. Then we describe the set of extracted features used in polarity classification of French movie reviews. Finally we provide and discuss the obtained experiment results and end up by drawing some conclusions and ideas for future work.

2 Previous Work

Classical approaches in text retrieval and categorization has so far focused on mining and analyzing factual information such as entities, events and their properties. They basically utilize Natural Language Processing methods and techniques in order to extract objective features aiding the classification and categorization of textual expressions [PLV02] with special emphasis on linguistic features in order to increase the performance. As linguistic features, [Gam04, MTO05] present syntactically motivated features, most of them based on dependency path information and modeled as high n-grams. Further linguistic features such as part of speech, negation, verbs modality, and semantic information (from WordNet for instance) are recently explored [WK09, TNKS09].

Much of the previous work focuses on defining the characteristics of conveyed opinions on the basis of textual data with processing granularity ranging from words, to expressions, sentences and documents. For this purpose, statistical approaches have been coupled with semantic approaches in order to detect opinions and sentiments in textual spans with different compositional structure [KH04, WWH05]. Semantic approaches aim at classifying sentiment polarity conveyed by textual data using commonsense, sentiment resources, as well as linguistic information. For instance, [HL04, ES05, NSS07, Den08] classify polarity using emotion words and semantic relations from WordNet, WordNet Gloss, WordNet-Affect and SentiWordNet respectively.

An important theoretical issue in the semantic approach is still how to define the semantic orientation of a word in its context. Some studies showed that restricting features to those adjectives would improve performance. [HM97] have focused on defining the polarity of adjectives using indirect information collected from a large corpus. However, more researches showed that most of the adjectives and adverbs, a small group of nouns and verbs possess semantic orientation [ES05, MTO05, TL03].

Only very few work [Den08, ACS08] have explored sentiment analysis in a multilingual framework such as Arabic, Chinese, English, German and Japanese. Their methodology is based on standard translation from target language to English in order to reuse existing English corpora and resources for polarity classification.

3 Feature Design

We have defined three categories of features: lexical, morpho-syntactic and semantic features. Lexical and morpho-syntactic features have been formulated at the word level, whereas semantic features have been formulated at the review level[1].

3.1 Lexical Features

This is the baseline of our experiments and is mainly composed of word unigrams. The global assumption in this choice is that we tend to find certain words in positive reviews and others in negative ones. Each unigram feature formulates a binary value indicating the presence or the absence of the corresponding word at the review level.

Lemmatization is argued to be relevant in sentiment analysis in order to group all inflected forms of a word in a single term feature, especially French is a inflected language. For example the words *aimé*, *aimait* and *aimer* share the same polarity but will be considered as five separate features during the classification. When applying lemmatization, we would obtain a unique feature. Features reduction would improve the tuning of the training process.

3.2 Morpho-syntactic Features

Some studies showed that restricting features to specific part-of-speech (POS) categories, for instance adjectives would improve performance [HM97]. In our

[1] In supervised learning, a training and test corpus is first annotated. A learning algorithm is then applied to induce the training model on the basis of the selected features. As it is shown in [ES05, PLV02] for instance, probabilistic algorithms (Bayes, maximum entropy) and linear discrimination (Support Vector Machine) are the most appropriate for the task of polarity classification.

approach, POS tags are proposed to be used to enrich unigrams features with morpho-syntactic information so as to disambiguate words that share the same spelling but not the same polarity. For example, it would distinguish the different usages of the word *négatif* that can either be a neutral noun *un négatif* or a negative adjective *un commentaire négatif.* Moreover POS tags are useful to handle negation and to aid word sense disambiguation before polarity extraction in SentiWordNet as it will be detailed hereafter.

Negation is handled at the shallow level of morpho-syntactic constituency of sentences avoiding the heavy processing of its deep syntactic structure. The detection of negated forms is performed by searching specific patterns formed from the abundantly utilized lexicalized forms of negation combined with particular n-grams of POS categories. We defined two simple patterns that cope with the negation form (1) at the verb level for example *le scénario* ***ne*** *brille* ***pas*** and (2) at the adjective and noun level for example ***sans*** *histoire originale.*

The scope of the negation is fixed with respect to a predefined context window of n POS categories within a textual span limited by a punctuation sign. We invert the polarity of the n verbs, nouns and adjectives within the context of each detected negation. We do not cope with other composed forms of negation such as conditional, double negation, the counter factual subordinates and modalities. The entailments of such a polarity inversion are first situated at the lexical level; unigrams features are inverted during features vector construction that is if we consider the previous example, in stead of having in the feature *original*, we would have a separate feature *!original* in the vector; second at the semantic level, polarity is inverted from positive to negative and vice-versa in the calculation of the overall polarity of a review as we will detail in the following section.

3.3 Semantic Features

As it is shown is previous work [HL04, ES05, NSS07, Den08], the incorporation of corpus and dictionary based resources such as WordNetAffect, SentiWordNet and Whissell's Dictionary of Affect Language contributes in improving the sentiment classification. Based on such results, we use the lexical resource SentiWordNet[2] to extract word polarity and calculate the overall polarity score of the review for each POS tag. SentiWordNet is a corpus-based lexical ressource constructed from the perspective of WordNet. It focuses on describing sentiment attributes of lexical entries describe by their POS tag and assigns to each synset of WordNet three sentiment scores: positivity, negativity and objectivity.

Since SentiWordNet describes English lexical resources, we go through a word-translation from French to English before polarity extraction. Words are lemmatized before being passed through the bilingual dictionary. We use POS information as well as the most frequently[3] used sense selection to disambiguate senses and predict

[2] SentiWordNet 1.0.1.

[3] This choice is based on the assumption that reviewers spontaneously use an everyday language.

the right synset. We only considered the positivity and the negativity features for the four POS tags noun, adjective, verb and adverb for this task.

More specifically, we added for each review and for each POS tag two features holding the scores of negativity and positivity as extracted from SentiWordNet. These two scores are calculated as the sum of polarities over all the words of the review respecting POS categorization. For example, for a given review, we obtain the following semantic features vector *(neg_adv, 6.38); (pos_adv, 1.25); (neg_noun, 0.12); (pos_noun, 0.50); (neg_adj, 0.62); (pos_adj, 0.12); (neg_verb, 0.12); (pos_verb, 0.38).*

4 Experiments

Since we didn't find any available sets of annotated data (already classified as negative or positive) of French movie reviews, we collected our own data from the web[4]. We extracted a corpus of 2000 French movie reviews, 1000 positive and 1000 negative, from 10 movies, 1600 were used for training and 400 for testing. We included reviews having a size between 500 and 1000 characters.

Prior classification of the corpus is elaborated according to user scoring: positive reviews are marked between 2.5 and 4 whereas negative reviews are marked between 0 and 1.5[5]. This prior classification is based on the assumption that the scoring is correlated to the sentiment of the review.

For our experiments, the data was preprocessed with the TreeTagger[Sch94], a French POS tagger and lemmatization tool. We applied Support Vector Machine (SVM) classification method and utilized SVM^{Light} [Joa98] classification tool with its standard configuration (inductive classification using linear kernel function)[6] to implement a series of experiments where each time we define a set of combined features and evaluate the accuracy of the approach. The simple validation method (data set division into traning and test corpora) has been applied to evaluate the approach.

4.1 Results and Discussion

The results of the following experiments are summarized in Table 1 below. For each experiment labeled from (1) to (9), we present the number of used features and the accuracy measured on the test corpus.

[4] We extracted spectators reviews from http://www.allocine.com

[5] Scores are bounded between 0 (for very bad) and 4 (excellent) with a step of 0.5. Reviews scored with 2 are not considered in the construction of our corpus since it is hard to manually classify them as positive or negative opinions.

[6] SVMLight software and detailed descriptions of all its parameters are available at http://svmlight.joachims.org.

Table 1 Performance of most relevant feature sets

Features	# of features	Results [%]		
		Pos.	Neg.	Global
(1) Unigrams	14635	92.00	91.00	91.50
(2) Unigrams + lemmatization	10624	92.00	93.00	92.50
(3) **Unigrams + lemmatization + negation**	**12002**	**92.50**	**94.00**	**93.25**
(4) Unigrams + lemmtization. + POS	12229	93.00	92.50	92.75
(5) Unigrams + lemmatization + POS + negation	13625	92.50	93.50	93.00
(6) Unigrams + lemmatization + POS (ADJ)	2109	79.50	92.00	85.75
(7) Unigrams + lemmatization + POS (ADJ) + negation	2492	80.00	91.00	85.50
(8) **Unigrams + lemmatization + polarity**	**10632**	**93.00**	**93.50**	**93.25**
(9) Unigrams + lemmatization + negation + polarity	12010	93.00	92.50	92.75

4.1.1 Lexical Features

Similarly to [PLV02] we encoded all words features as binary values indicating the presence or the absence of a word in a review. As a first step, we included the entire set of words without applying any specific filtration method.

The accuracy in experiment (1) using the entire set of words is found 91.50%; when comparing this result to Pang et al. [PLV02] who reported an accuracy of 82.90% on English movie reviews using similar features, we find that our results are approximately 10% higher. We believe that this gap is due to the nature of our corpus and the size of our reviews (the collected French reviews are shorter). Moreover, the incorporation of the lemmatization process (2) increases the accuracy by 1.00% up to 92.50%. This was quite expected since French is an inflected language. In experiment (3) we find that negation, although it is processed in a simple form, improves the results to reach 93.25%. Moreover, we notice that the classification of the negative reviews is being improved by the negation processing (from 93% up to 94%) which noticeably means that negation is relatively efficient at this lexical level.

After looking deeply through the reviews, we found that misclassification is mainly due to the following difficulties.

Misspellings. Misspelled words are not standard unigrams and hence could not regularly be present in the training data. Reviews containing a large number of misspellings would have their features significantly reduced and so provide very poor information for the classification. We noted that isolated and common misspellings don't affect much the classification but reviews which contain relatively many misspellings tend to be misclassified. Sometimes misspellings are hard to be automatically corrected, especially those made voluntary in order to express a kind of stress and emphasis such as *énnnnorme*. The problem with such kind of words is that they are irregular in the corpus. For example, *énnnnorme* is highly positive but it is not present in the feature set so it is not useful. Quite misspelled reviews tend to be misclassified.

Neutral and mixed reviews. Reviews manually interpreted as neutral such as *le film est visuellement réussi mais le scénario est d'une banalité affligente* are randomly classified according to the dominant sentiment of contained words. As a matter of fact, reviewers tend to argue their opinion by posting simultaneously positive and negative arguments organized in a concession or a contrast rhetorical form. Lexical classification shows its limits when the abundant polarity of text spans is not coherent to the final retained opinion, typically the case of a reviewer who starts by verbosely listing the film drawbacks and ends by confessing his admiration and concisely posting his favorable judgment. A further difficulty concerns ironic expressions such as *trop fort les gars* that has a negative polarity although it is composed of positive words. In addition, the classical issue of idiomatic expressions, proverbs and sayings could in some cases have a polarity that doesn't follow the polarity of its composed words, and hence affect negatively the classification.

4.1.2 Morpho-syntactic Features

In further experiments, we appended POS tags to every lemmatized unigram so as to disambiguate same unigrams having different syntactic roles. However, the effect of this information seems to be not quite relevant, as depicted on line (4) of Table 1, the accuracy is only increased by approximately 0.25% up to 92.75%. When applying the negation processing (5) to the same experiment, results were slightly improved (up to 93.00) but still not higher than experiment (3) where no POS tags were used. This entails that ambiguity at the morpho-syntactic level of the reviews does not much effect on the polarity classification. Thus, we eliminate this feature from our next experiments.

When restricting unigrams features to only adjectives (6), the performance is getting worse; accuracy is decreased by 6.75% down to 85.75% comparing to (2) and the feature set is reduced by approximately 80%. In order to understand such inconsistency, we look deeper at the accuracy of positive and negative reviews separately. On a one hand, we notice that negative reviews are better classified than positive ones. On the other hand, we have found, in additional experiments, that negative reviews contain relatively an important number of positive adjectives (generally in the negative form). In the first experiment (6) and before processing the negation, these positive adjectives are assumed to negative features in the training model, which induces a further difficulty when classifying positive reviews containing these positive adjectives. However, in the second experiment (7), even after negation processing, the results didn't improve which obviously entails that the scope of processed negation didn't capture the adjectives and was mostly local to the verbal phrase. This last experiment is in contradiction with the results of [HM97] but confirms the results of [PLV02].

As we have already described the negation processing in the previous section, the most used form of negation is that detected at the verbal phrase level. Since deep syntactic dependency analysis of reviews is a quite costly task, it is difficult in this case to capture the adjectives related to the negated verb. Heuristic rules discussed previously such defining a context of a bag of words after the negated verb is not likely to give satisfactory results.

4.1.3 Semantic Features

A part from the lexical and the POS features, we extend in our experiments the features set to words polarity extracted from SentiWordNet and formulated as a score representing the overall negativity and positivity of words in the reviews. As shown on the table 1 experience (8), results are improved by 1.75% up to 93.25% compared to lemmatized unigrams experiment (2). The main reason of such an barely perceptible improvement is the failure of extracting polarity information of words from SentiWordNet: among 2000 adjectives, we got the polarity information of only 800 entries in SentiWordNet (40% of success). This extraction problem is mainly due to the following problems.

Translation errors. We translate words from French to English so as to cope with SentiWordNet interface. However, the quality of translation significantly affects the results of semantic polarity extraction; this is mainly due to the following reasons.

- The bilingual translator doesn't preserve the POS of words. For example, the *noun méchant* is translated into *wicked* which is implicitly an *adjective* and not a *noun*. Since the translator does not reveal information about the POS change after translation, *wicked* is assumed to be a *noun*. However, the *noun wicked* doesn't exist in SentiWordNet.
- Moreover, even if the translation is correct, it happens that the parallel words do not share the same semantic orientation across both languages due to a difference in common usage, for instance the French *positive* adjective *féériques* is translated into the *negative* English adjective *magical*

Lemmatization and POS tagging errors. Misspellings are not standard unigrams and hence could not be found in SentiWordNet. Reviews containing a large number of misspellings would have their overall polarity uncorrect. In addition, misspellings and other lexical errors (for example punctuation, use of parenthesis *permanente(c'est* and composed words *as-tu-vu*) could significantly affect the results of lemmatization and POS tagging tasks elaborated by TreeTagger. In fact, TreeTagger is not implemented to cope with everyday French language as found in spontaneous movie reviews.

Negation. As shown in the last experiment (9), the integration of negation processing didn't improve the results. For reminder, the negated verbs, nouns and adjectives would have their extracted polarity score from SentiWordNet inverted. We explain such an outcome by two reasons (i) negation didn't capture properly adjectives (considered as the most subjective lexicon) (ii) bilingual translation of subjective lexicon was not very precise.

5 Conclusions

In this paper, a supervised approach to sentiment analysis of French movie reviews in a bilingual framework was described. The simple validation method (data set division into traning and test corpora) has been applied to evaluate the approach. Preliminary results have shown that the combination of lexical, morpho-syntactic and semantic features achieves relatively good performance in classifying French movie reviews according to their sentiment polarity (positive, negative). Several problems having an effect upon the results of the classification were highlighted and potential solutions were discussed.

In order to extract the semantic orientation of words from SentiWordNet, we went through a standard word-translation process. Although translation does not necessary preserve the semantic orientation of words due to the variation of language common usage especially when it comes to spontaneous reviews on the web, and in spite of all its side effects, it has been argued that dictionary-based approach could contribute to achieve better results. Even if our first experiments showed little significance, further improvements have been proposed accordingly, particularly concerning negation processing.

In future work, the method will be analyzed within a larger training and test sets and evaluated using more sound methods such us cross validation. Further shallow linguistic analysis will be elaborated such as misspelling correction, more elaborated negation processing, WSD and elimination of out of scope text spans from reviews, in addition to the improvement of the translation task using French-English EuroWordNet.

References

[ACS08] Abbasi, A., Chen, H., Salem, A.: Sentiment analysis in multiple languages: Feature selection for opinion classification in web forums. ACM Transactions on Information Systems (TOIS), Article 12 26(3) (June 2008)

[Den08] Denecke, K.: Using sentiwordnet for multilingual sentiment analysis. In: Proceedings of the IEEE International Conference on Data Engineering (ICDE 2008), pp. 507–512 (2008)

[ES05] Esuli, A., Sebastiani, F.: Determining the semantic orientation of terms through gloss classification. In: Proceedings of CIKM 2005, pp. 617–624 (2005)

[ES06] Esuli, A., Sebastiani, F.: Sentiwordnet: a publicly available lexical resource for opinion mining. In: Proceedings of the 5th Conference on Language Resources and Evaluation LREC, vol. 6 (2006)

[Gam04] Gamon, M.: Sentiment classification on customer feedback data: Noisy data, large feature vectors, and the role of linguistic analysis. In: Proceedings of the 20th International Conference on Computational Linguistics, pp. 611–617 (August 2004)

[HL04] Hu, M., Liu, B.: Mining and summarizing customer reviews. In: Proceedings of Knowledge Discovery and Data Mining, KDD 2004 (2004)

[HM97] Hatzivassiloglou, V., McKeown, K.R.: Predicting the semantic orientation of adjectives. In: Proceedings of the 8th Conference on European Chapter of the Association for Computational Linguistics, pp. 174–181 (1997)

[HR10] Hassan, A., Radev, D.: Identifying text polarity using random walks. In: Proceedings of the 48th Annual Meeting of the Association for Computational Linguistics, pp. 395–403 (2010)

[Joa98] Joachims, T.: Making large-scale svm learning practical. ACM Transactions on Information Systems, TOIS (1998)

[KH04] Kim, S.-M., Hovy, E.: Determining the sentiment of opinions. In: Proceedings of the 20th International Conference on Computational Linguistics (COLING 2004), pp. 1367–1373 (August 2004)

[MTO05] Matsumoto, S., Takamura, H., Okumura, M.: Sentiment classification using word sub-sequences and dependency sub-trees. In: Ho, T.-B., Cheung, D., Liu, H. (eds.) PAKDD 2005. LNCS (LNAI), vol. 3518, pp. 301–311. Springer, Heidelberg (2005)

[NSS07] Nastase, V., Sokolova, M., Shirabad, J.S.: Do happy words sound happy? a study of the relation between form and meaning for english words expressing emotions. In: Proceedings of Recent Advances in Natural Language Processing (RANLP 2007), pp. 406–410 (2007)

[PLV02] Pang, B., Lee, L., Vaithyanathan, S.: Thumbs up? sentiment classification using machine learning techniques. In: Proceedings of the Conference on Empirical Methods in Natural Language Processing (EMNLP), pp. 79–86 (July 2002)

[Sch94] Schmid, H.: Probabilistic part-of-speech tagging using decision trees. In: Proceedings of the International Conference on New Methods in Language Processing, pp. 44–49 (1994)

[SV04] Strapparava, C., Valitutti, A.: Wordnet-affect: an affective extension of wordnet. In: Proceedings of the 4th International Conference on Language Resources and Evaluation (LREC 2004), pp. 1083–1086 (May 2004)

[TL03] Turney, P.D., Littman, M.L.: Measuring praise and criticism: Inference of semantic orientation from association. ACM Transactions on Information Systems (TOIS), 15–346 (2003)

[TNKS09] Thet, T.T., Na, J.-C., Khoo, C., Shakthikumar, S.: Sentiment analysis of movie reviews on discussion boards using a linguistic approach. In: Proceedings of the 1st International CIKM Workshop on Topic-Sentiment Analysis for Mass Opinion Measurement (2009)

[Tur02] Turney, P.D.: Thumbs up or thumbs down? semantic orientation applied to unsupervised classification of reviews. In: Proceedings of the 40th Annual Meeting of the Association for Computational Linguistics (ACL 2002), pp. 417–424 (2002)

[WK09] Wiegand, M., Klakow, D.: The role of knowledge-based features in polarity classification at sentence level. In: Proceedings of the Florida Artificial Intelligence Research Society Conference (FLAIRS Conference 2009) (2009)

[WK10] Wiegand, M., Klakow, D.: Bootstrapping supervised machine-learning polarity classifiers with rule-based classification. In: Proceedings 1st Workshop on Computational Approaches to Subjectivity and Sentiment Analysis, WASSA (2010)

[WWH05] Wilson, T., Wiebe, J., Hoffmann, P.: Recognizing contextual polarity in phrase-level sentiment analysis. In: Proceedings of the Conference on Empirical Methods in Natural Language Processing (EMNLP 2005), pp. 347–354 (October 2005)

Querying over Heterogeneous and Distributed Data Sources

Maria Sokhn, Elena Mugellini, and Omar Abou Khaled

Abstract. The current web integrates diverse sources of heterogeneous and distributed data (XML database, relational database, Peer database within a P2P network etc.). Integrating data from multiple heterogeneous sources leads to the coexistence of different data models and consequently different query languages. Hence, there is a strong need to address the issue of handling queries over heterogeneous and distributed data. The structural and semantic heterogeneity of data makes the development of custom solutions for querying a time-consuming and complex task. In this paper we propose a query engine system, *Virtual-Q*. The system is based on data semantic categorization concept which assigns levels of importance according to the semantic level of the data sources. This approach is referred in our paper as the ontology-based approach. The *Virtual-Q* system aims at providing users a simple and transparent information access. This paper presents also a preliminary prototype validating the proposed architecture.

Keywords: Heterogeneous Data Sources, SemanticWeb, Virtual Query.

1 Introduction

The semantic web technologies have brought to users the possibility to semantically enrich the data. In the recent years, the ontology, one of the most known technology in the semantic web, has been a widely used technique in the WWW. However this situation raises the issue of heterogeneous data integration, management and retrieval. Integrating data from multiple heterogeneous sources leads to the coexistence of different data models and scheme and therefore different query languages. Hence, there is a strong need to address the issue of handling queries over

Maria Sokhn · Elena Mugellini · Omar Abou Khaled
University of Applied Sciences of Western Switzerland,
Fribourg, Boulevard Perolles, 80, 1700, Fribourg
e-mail: maria.sokhn@hefr.ch, elena.mugellini@hefr.ch,
omar.aboukhaled@hefr.ch

E. Mugellini et al. (Eds.): Advances in Intelligent Web Mastering - 3, AISC 86, pp. 29–38.
springerlink.com

heterogeneous and distributed data. The structural and semantic heterogeneity of data makes the development of custom solutions for querying these data time-consuming and complex. This issue can be divided into three main parts: dispatching important search information depending on the data sources, creating database-specific queries and merging results from several sources. The system *Virtual-Q* we present in this paper, is a system developed within the framework presented in the paper [1]. *Virtual-Q* is based on a noval virtual query engine architecture. The aim of our system is to provide transparent and easy access for end-users to retrieve data from heterogeneous sources. By transparent, we mean that the user should be as less as possible involved in technical issues. Indeed, aspects of configuration (model extraction, schemes merging, etc.) for example could be obstacles for some users not familiar with.

This paper is organized as follows: Section 2, presents a brief overview of the most significant related works. Section 3, introduces the *Virtual-Q* system. Section 4, presents the preliminary prototype. Finally section 5 concludes the paper and presents the future steps.

2 Related Works

The related works are grouped into three parts: (I) Frameworks and systems, (II) the heterogeneous data sources integration and (III) sub-queries reformulation.

I- Frameworks and systems. The SemanticLIFE [8] project of Vienna University of Technology [18] attempts to get closer to the vision of Memex of Vanevar Bush [13]. As part of this project, a virtual query system and a virtual query language have been developed. The system is fed with data and external sources can be queried too. It is based on the ontology to remove ambiguities occurring in the user queries. The semantic integration of heterogeneous data is the main theme of the SIRUP [16] project lead by the University of Zurich [17]. A semantic multidatasource language is used to declaratively manipulate the so-called IConcepts. They provide explicit, queryable semantics by connecting IConcepts to concepts of ontologies. Another framework is the European project NEPOMUK [10], which integrates the Gnowsis Semantic Desktop [9] which aims to add the notion of web semantic to classical desktop applications. The data are integrated in a desktop web server using adapters and the user can navigate through the documents. The MIT [14] has investigated new approaches for a user to manage information using relationship or anything else. They have developed several applications within the Haystack Project [15]. Each of the described system has a specific approach that can be categorized into two main approaches: (1) Either the data are directly fed into the system or (2) the system accesses each time the sources to retrieve data. We can find hybrid solution as well. The *Virtual-Q* system is based on the second concept. The engine accesses the data sources every time a query is raised. This approach has the advantage of keeping the data up-to-date. However, an effort is required to merge the resulting data.

II- Heterogeneous data sources integration. The frameworks and the systems presented above require a common concept to achieve their goals: querying heterogeneous data sources. Several papers have already raised this issue. [2] and [3] both propose architectures with a global merged ontology/taxonomy. Each source has its own local ontology/taxonomy. [5] describes the SINGAPORE architecture which knows similarities/conflicts between sources schemes and rules defined manually by an administrator. [7] introduces the notions of Concepts, Properties and Categories. Then, when a new source is registered, a mapping between the source and the supported notions is made. [4] presents XLive, an XML Light Integration Virtual Engine. Here, the structures of all connected sources are simply written down in an XML configuration file. [6] exposes the TSIMMIS project. It uses a schema-less approach which is well suited for sources with dynamic contents. It does not use global schema describing the sources or fixed schema for the data. This approach is made possible through a hierarchy of mediators. There are three main concepts for the integration of heterogeneous data sources: **(1)** Terms mapping between local and global schemes, taxonomies, ontologies or other concepts, **(2)** the structure of the sources are explicitly described and **(3)** use of predefined/template queries to lead to the concerned sources. All the heterogeneous data source integration systems share a common approach referred in this paper as the *fixed point*. *fixed point* means that a part of data is explicitly filled by users. Indeed, there is either a mapping to do, a global schema to build or a source structure to describe. The *Virtual-Q* system aims at reducing the technical involvement of users. To achieve this goal, the engine must be aware of its data and be able to automatically perform some tasks, one of which is the reformulation of source-specific queries which is the so called sub-queries, detailed in the following section.

III- Sub-Queries Reformulation. Different approaches exist to handle the creation of source-specific queries. These solutions depends on the way the data sources are integrated. In most of the cases, the transformation from global query to local queries is done with the help of the global-local mapping or with relations in global or merged ontologies [2, 7, 3]. In a similar way, [5] specifies the conflicts/similarities between schema and rules for correspondences. Generally, such solutions are made for relational or equivalent data sources. The TSIMMIS [6] project has predefined queries that delimit in same time the sources' capabilities. Every user's query is matched with them using Pattern Matching techniques and their stored query plans are executed. On the other side, the SIMS system needs complex input query written in the LOOM language. As the information sources' schema, the global domain model and relations between the both are as well described with LOOM, the reformulation is done through operators and LIM (Loom Interface Manager). Several source-queries are possible but the integration of a query planning allows choosing one of them. The paper [11] presents methods to divide a global query into sub-queries. A global ontology is mapped with local information source (local mapping) and altogether they form the global mapping. As usual, the local results are integrated as result of the global query. The algorithm is working under simple mapping compositions but they announce future work for complex mapping. There are only two types of solutions proposed in the reviewed papers: using mapping information to match fields' names or predefined queries. Unfortunately, no paper on *real* creation of sub-queries has been found. In general, the paper stops when arriving at this point [5, 2].

VI- Discussion. This state of the art stated below confirm the existence of several frameworks and systems that have been developed. These solutions designed for heterogeneous data sources integration present similar approach, yet as far as we know no real proposition has been made for a complete sub-queries formulation. Every solution based its design on a fixed point helping the retrieval and integration of data sources. The next section presents our novel approach which aims at removing this fixed point. We propose a virtual query architecture where the engine is able to analyze any type of query. Based on this analyze, source-specific sub-queries are created, helping retrieving data from heterogeneous sources. Thus users are hardly involved in technical issues.

3 Virtual-Q System

As described above existing solutions base their work on the *fixed point*. In this paper we propose the *Virtual-Q* system, a novel approach based on an innovative virtual query engine. This new architecture integrates concepts as query analysis or sub-queries formulation, which facilitate a transparent access to heterogeneous data. Existing architectures are often based on existing elements that simplify the issue of sub-queries reformulation: a global ontology and local ontology for each data source, using only XQuery and working on XML data with metadata describing sources' structure or an administrator determines similarities and conflicts between sources and defines rules for inter-schema correspondences. *Virtual-Q* aims to provide mainly an easy-to-use system by avoiding the complex administration work. Therefore we designed an autonomous Query Engine able to query heterogeneous data sources without knowing in advance the structure by the user/administrator. The user takes advantage of this approach. He easily adds or removes data sources at any time. The query are free, a text field, and can be based on theme/topics to improve the results by searching only in the pertinent data sources. No knowledge of data source structure is requested, but sometimes, connection information is necessary (e.g. url connection, username and password). From the users' point of view, there are only advantages because the Virtual Query Engine attempts to reduce his implication in the data sources integration. In contrast, the task of the developers is now more complex. They have to assume the formulation of source-specific sub-queries, which is not a trivial job according to the kind of sources. The free query is therefore analyzed as well.

3.1 Virtual Query Engine Architecture

The global system (Figure 1), is divided into four main parts: **(1)** the user interface, **(2)** the Virtual Query Engine (VQE), **(3)** the external data sources and **(4)** the external reasoning models. The user interface displays the information for the user and proposes an interface to formulate the queries. The external data sources are the heterogeneous sources the engine will query. The external reasoning models are sources that could help the engine when it reasons on the results. Lastly, the Virtual Query Engine is the core of the global system which is discussed in this paper. The Figure 2 details the architecture of the Virtual Query Engine. The Virtual Query Engine is composed of several modules. The modules that process the

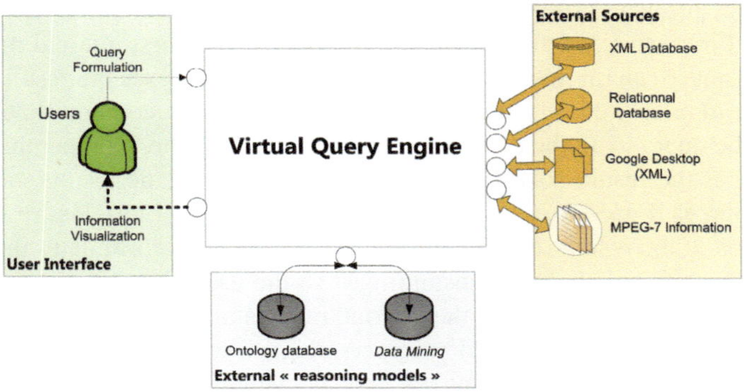

Fig. 1 Virtual-Q system overview

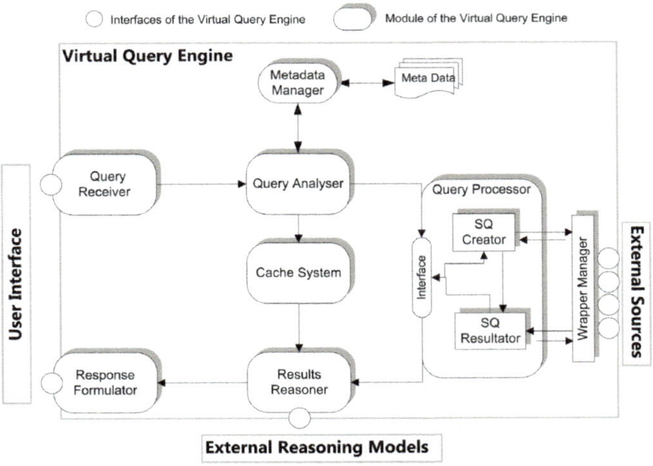

Fig. 2 Virtual Query Engine architecture

query are linked together and form a chain. This modularity allows to clearly define the task(s) of each module. We also can easily change a module or add a new one. Finally, several modules could be chained to do a specific task. Moreover, two modules are responsible for managing the internal "configuration" of the engine and the wrappers (especially forwarding the query to the right wrappers, thus to the right data sources). The different modules composing the Virtual Query Engine and their behaviors are described hereunder:

Query process module: (1) Query Receiver: receives the original user query and transforms it in the format used inside the Virtual Query Engine. We talk here about User Query. **(2)** Query Analyzer: analyses the free query to find pertinent data that could be used later in the process. Therefore, it retrieves first the best schema trough the Metadata Manager and the Wrapper Manager. Then, with different methods analyses the query and stores those data in the query. We talk about Virtual Query. **(3)** Query Processor: contains two internal modules. The first internal module creates a first query run on the best data source, possibly analyses the results for additional helpful data and transmits the Virtual Query to the second module. That one forwards the query to all the concerned wrappers through the Wrapper manager and lastly merges the results. **(4)** Result Reasoner: reasons on the results, helped with external sources if necessary, to improve them: suppression of useless results, better organization of the results, etc. **(5)** Response Formulator: reformulates the response into a standard format before transmitting it to the user. **Resource management: (1)** Metadata Manager: manages the internal metadata of the Virtual Query Engine (connected data sources, installed wrappers, themes with source ranking and reasoning sources). **(2)** Wrapper Manager: is responsible for forwarding the queries to the concerned wrapper(s).

3.2 Query Process

Once the query is transmitted by the user interface to the query engine, it is transformed into a format allowing the addition of extra data. This query passes through the chain of modules. Each module is free to use these data or to add additional. In principle, the data are added during the analysis phase and used during the next phases like sub-queries reformulation or results reasoning. The query process (Fig. 3) acts as follows:

(1) The user enters the query and transmits it to the engine. **(2)** The query is analyzed, using the best data source schema if necessary. **(3)** The query is reformulate in source-specific queries, which are executed on the corresponding sources. The query is first run on the best source in order to find more helpful data for sub-queries reformulation and then run on the other sources. **(4)** The engine merges the results and reasons on them to remove useless or organize them for example. **(5)** The results are formatted in the correct output format. **(6)** Finally, the results are displayed to the user. **Query Analysis** This section details the query analysis and sub-queries reformulation concepts. The query analysis allows finding

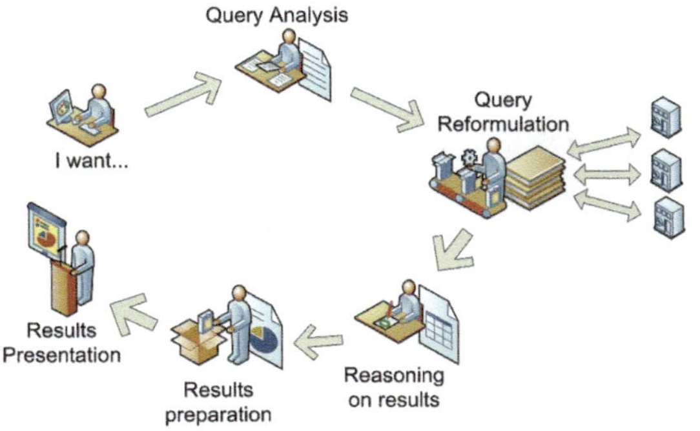

Fig. 3 Query process approach

pertinent information for users queries. These data are then be used along the query process. In order to improve the query analysis, we suggest the following approach: **(1)** Data sources schemes: Each data stored into a source is compliant to a schema or model, even the most basic one. According to the kind of source, this schema will be more or less complex and complete. The idea is to request this schema to the data source. Normally, the most data sources allow retrieving their schemes or models. Sometimes, it can also be known differently. **(2)** Ranking of the schemes: One knows as well that some schemes are stronger than others. For example ontology provides more structured and rich information than then Google Desktop XML Schema. So, the second idea is to rank the available schemes for each theme and to store it in the metadata. Like this, the VQE knows the best schema for a given theme. **(3)** Using the schema: Once we have the best schema or model, we can use it to analyze the queries. This could be done exploiting the fields names or types for example. Moreover, since we know the best schema, we can refine the creation of sub-queries using the results of the query on the best data source. Secondly, it is important to know which information could be "produced" from the query and how. Some possibilities are: relevant data sources (using metadata and source ranking), Keywords (fields' names of data sources, lists, ontologies, etc.), stop words (Black list, database, etc.), summary (who, what, where, etc.), boolean functions (predefine keywords or symbols: OR, AND, &&, etc.), relations between terms (ontologies, etc.). In most cases users only write few keywords for their query, consequently some of the methods presented above are no more adequate. Sub-queries formulation presents a key-solution for this issue.

Sub-Queries Reformulation. The query reformulation is one of the main challenges of *Virtual-Q*. Indeed the integration of heterogeneous data sources and the fact that the user asks free queries add the constraint of transforming the original query into source-specific queries. This means that for the query *Greece*, the correct SQL query is *Select X From Y Where Z* for a relational database or XQuery *Full-text search* query for XML documents must be generated. Moreover, the approach chosen by *Virtual-Q* (no global schema and free query) increases the difficultly of the query reformulation. The analysis module gets round the free queries creating *structured queries*. Then the sub-queries are composed thanks to this structured information. Concerning the global schema, *Virtual-Q* does not have an established one, we consider that the module is self-learned based and therefore it keeps track of links between keywords, local information and between local schemes. The complexity in developing a wrapper depends on the kind of source that is being handled. Wrappers could be *Full-text Search*, *API-based* or *fields-based*.

4 Prototype

In order to validate the new concept proposed in this paper, a preliminary prototype has been developed in Java 1.6. Currently, bases of the query process have been implemented as well as the Wrapper Manager and the Metadata Manager. Two wrappers have also been created: one for Google Search Desktop through an API and another for searching inside MPEG-7 files (an XML based language for multimedia contents describtion). The graphical interface is composed of two windows. The user window allows running a query and the administration window (Fig. 4) to manage the wrappers and data sources. It also displays the log of the engine. The other technical choices explained in the following relate to the data format for queries, the communication between the modules and the metadata management.

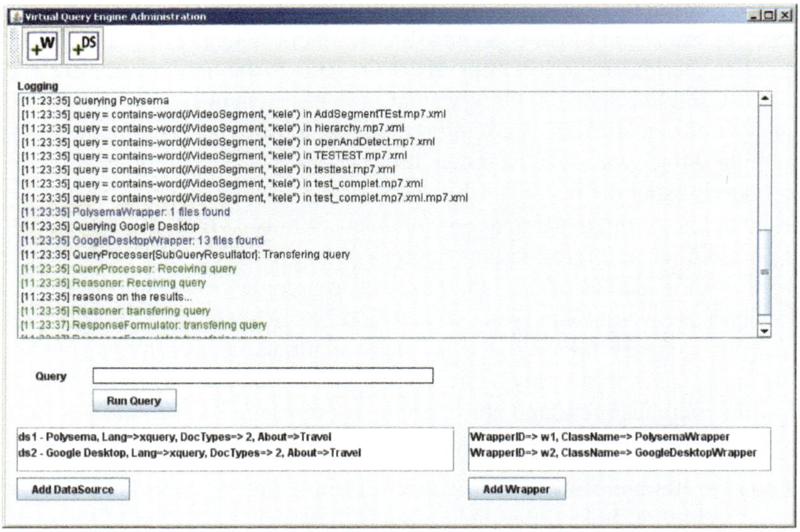

Fig. 4 *Virtual-Q* prototype: administration window

Query: The principal format used all along the process is XML. First, XML is an easy-readable format for which many tool already exist. Secondly, XML is the base for OWL and RDF. So, it is possible to easy translate data from on format into another when needed. Thirdly, XML allows envisaging external treatments about the query because the transmission of data is easy and even Web Services technologies could be used. In the engine, the XML data are encapsulated in Java Object. The "XML query" is compliant to an XML Schema but no validation is done because only the software is expected to modify these data.

Inter-modules communication: The second problem is the communication between the modules. The only data that is exchanged between them should be the query in the corresponding format. The *Source-Listener* solution has been chosen for the implementation. Each module is aware of which one wants to receive the query after having processed it. When the chain of modules is created, a module can add itself as a *listener* of another. Then the *source* of the event calls a specific method of each listener. For this reason, each listener must be conform with a common specification for all listeners.

Metadata: The Metadata are stored in an XML file. The access to the data is made using the JAXP API. However, like the Metadata Manager does not directly work on the raw data, it is still possible to change the storage type without affecting the whole engine. The data are compliant to an XML Schema but no validation is done because only the software is expected to modify the file.

5 Conclusion and Future Work

In this paper we have proposed a novel approach for the integration of heterogeneous data sources. The novelty of the *Virtual-Q* system is to propose a system that could be used by end-user with no specific technical knowledge. Indeed, our virtual query engine architecture attempts to offer a transparent access for users to retrieve data from heterogeneous data sources. The "data source semantic categorization" allows improving the *Virtual-Q* information retrieval by assigning importance levels to data sources according to their degree of "semantisation". A preliminary prototype, presented at the end, validates the proposed architecture. As next step, we plan, to improve the modules developed so far. Indeed, the query analysis and the sub-queries reformulation require additional work to be effective. An operational prototype should be soon available with new functionalities and improvements allowing us to proceed with a quantitative evaluation of ours results.

Acknowledgment. We would like to thank David Godel for his precious and valuable work within this project.

References

1. Maria, S., et al.: Knowledge management framework for conference video-recording retrieval. In: 21st International Conference on Software Engineering and Knowledge Engineering, Boston, pp. 709–714 (2009)
2. Munir, K., et al.: Semantic information retrieval from distributed heterogeneous data sources. In: FIT special track on Bioinformatics for Academia and Industry, Islamabad (2007)
3. Tzitzikas, Y., et al.: Mediators over taxonomy-based information sources. The VLDB Journal 14(1), 112–136 (2005)
4. Dang-Ngoc, T., et al.: Xlive: An xml light integration virtual engine. In: Demonstration at the 21eme Conference on Bases de Donnees Avances (BDA) Conference, Saint-Malo (2005)
5. Domenig, R., et al.: A query based approach for integrating heterogeneous data sources. In: 9th International Conference on Information and Knowledge Management, pp. 453–460. ACM, New York (2000)
6. Hammer, J., et al.: The tsimmis project Integration of heterogeneous information sources. In: IPSJ Conference, Tokyo, pp. 7–18 (1994)
7. Kai-uwe, S., et al.: Concept-based querying in mediator systems. The VLDB Journal 14(1), 97–111 (2005)
8. Huu Hoang, H., et al.: Towards a new approach for information retrieval in the semanticlife digital memory framework. In: International Conference on Web Intelligence, pp. 485–488. IEEE Computer Society, Washington (2006)
9. Sauermann, L.: The gnowsis-using semantic web technologies to build a semantic desktop. Diploma thesis, Technical University of Vienna (2003)
10. Sauermann, L., et al.: Overview and outlook on the semantic desktop. In: 1st Workshop on The Semantic Desktop ISWC, Galway (2005)
11. Jian, L., et al.: Query division and reformulation in ontology-based heterogeneous information integration. In: 15th International Conference on Computing, Mexico, pp. 186–196 (2006)
12. Ben Necib, C., et al.: Using ontologies for database query reformulation. In: 18th Conference on Advances in Databases and Information Systems (2004)
13. Memex, `http://www.theatlantic.com/past/docs/unbound/flashbks/computer/bushf.htm`
14. Massachusetts Institute of Technology, `http://www.mit.edu/`
15. Haystack Project, `http://groups.csail.mit.edu/haystack/`
16. SIRUP, `http://www.ifi.uzh.ch/dbtg/Projects/SIRUP/`
17. University of Zurich, `http://www.uzh.ch/`
18. Vienna University of Technology, `http://www.tuwien.ac.at/tu_vienna/`

Experiments in Bayesian Recommendation

Thomas Barnard and Adam Prügel-Bennett

Abstract. The performance of collaborative filtering recommender systems can suffer when data is sparse, for example in distributed situations. In addition popular algorithms such as memory-based collaborative filtering are rather ad-hoc, making principled improvements difficult. In this paper we focus on a simple recommender based on naïve Bayesian techniques, and explore two different methods of modelling probabilities. We find that a Gaussian model for rating behaviour works well, and with the addition of a Gaussian-Gamma prior it maintains good performance even when data is sparse.

Keywords: Recommender systems, Collaborative filtering, Bayesian methods, Naïve Bayes.

1 Introduction

Recommender systems are information filtering systems that are widely used on the web to suggest items to users based on their preferences. Collaborative filtering recommenders use item ratings to suggest items preferred by similar users, based on the assumption that people who have agreed in the past will agree in the future[10].

Recommendation accuracy suffers in situations where information is limited[7], such as in distributed[12] or context-aware[1] recommender systems. In addition, some of the more widely used recommender system algorithms such as memory-based collaborative filtering, are rather ad-hoc, and so it is difficult to make principled improvements. Motivated by these challenges, in this paper we present a simple recommender system based on probabilistic methods, which uses prior knowledge to reduce the impact of data sparsity.

Thomas Barnard · Adam Prügel-Bennett
Information: Signals, Images, Systems,
School of Electronics and Computer Science,
University of Southampton, SO17 1BJ, United Kingdom
e-mail: tcb08r@ecs.soton.ac.uk, apb@ecs.soton.ac.uk

E. Mugellini et al. (Eds.): Advances in Intelligent Web Mastering - 3, AISC 86, pp. 39–48.
springerlink.com

After looking at related work in Section 2, we look at making recommendations using Bayes' theorem in Section 4. We then present two models for modelling user ratings, the first based on a multinomial distribution in Section 5, and the second based on a Gaussian distribution in Section 6. Finally we present the results of our experiments in Section 7 before concluding in Section 8.

2 Related Work

Naïve Bayesian techniques have been used to produce recommendations before. Breese et al.[2] present a cluster model using a naïve Bayes classifier, which groups users based on their rating habits, before predicting ratings given cluster membership. Miyahara and Pazzani [9] present a recommender system based on a naïve Bayes model that makes binary rating predictions (i.e. like or dislike). This approach differs to Breese's as they create a separate model for each user. Wang et al.[13] present a probabilistic relevance ranking method that is similar to the item-based collaborative filtering algorithm[11]. They use a Beta distribution to add prior knowledge.

Our approach builds on the simple technique used by Miyahara and Pazzani [9], but we apply the technique to ratings on a numerical rather than binary scale. The naïve Bayes approach is often overlooked in favour of more complex models. In addition, we incorporate prior knowledge into our probability estimates, as Wang et al.[13] do with binary ratings.

3 Notation

Before we present our recommender it is necessary to define the notation we will use in the rest of this paper. We denote the set of all users in our recommender system by U, and the set of all items by I. The rating made by user u on item i is given by $r_{u,i} = k$, where $k \in \mathbf{K}$, the set of possible rating values. $r_u = k$ states that the user u rates any item with value k. The set of items for which user u has made a rating is given by I_u, and the set of users who have made a rating on item i is given by U_i. Finally we define the set of items for which two users u and u' have both provided a rating to be $I_{u,u'} = I_u \cap I_{u'}$.

4 Bayesian Recommendation

Bayes' theorem is a simple probabilistic technique that allows us to update our beliefs about the likelihood of an event occurring given the evidence. These techniques are said to be naïve, in that strong independence assumptions are made about the independence of features. Despite these assumptions naïve Bayes models can achieve good performance[4].

In the case of CF recommenders, our beliefs are the probability that a user will make a rating of a given class on an item, and our features are the ratings made by other users. To simplify calculations we consider priors and likelihoods to be independent of the item of interest, and incrementally update the posterior given each feature,

$$P(r_{u,i} = k | r_{u',i} = k') = \frac{P(r_u = k)P(r_{u'} = k' | r_u = k)}{\sum_{k''} P(r_u = k'')P(r_{u'} = k' | r_u = k'')} . \tag{1}$$

Posterior probabilities are combined to find the expected value of the rating $E(r_{u,i})$ as in [2],

$$E(r_{u,i}) = \sum_{k \in \mathbf{K}} P(r_{u,i} = k)k . \tag{2}$$

To estimate the priors and likelihoods we first take simple point estimates. Then we use all of our data to create Bayesian priors we can update with more specific information. These priors can be learnt on a server with access to large amounts of information before being passed to devices with more modest resources for distributed recommendation. In the following sections we investigate two different distributions for modelling probabilities using this method.

5 Multinomial Model

The simplest method of obtaining estimates for our priors and likelihoods is by normalising rating counts, which is equivalent to taking maximum likelihood estimates of the parameters of a multinomial distribution,

$$P(r_u = k) = \frac{n_{u,k}}{\sum_{k'} n_{u,k'}} , \tag{3}$$

$$P(r_u = k, r_{u'} = k') = \frac{n_{u,u',k,k'}}{\sum_{k''} \sum_{k'''} n_{u,u',k'',k'''}} , \tag{4}$$

where $n_{u,k}$ is the number of times user u has given an item a rating k, and $n_{u,u',k,k'}$ is the number of times user u has rated k, when user u' rated k'. Given these probabilities calculating the likelihood is trivial.

Where rating counts are zero, these probabilities will be zero, so to remove these zeros we apply Laplace smoothing, adding one to each of our rating counts. This is the model used by Miyahara and Pazzani[9].

5.1 Dirichlet Prior

Prior knowledge is incorporated into our multinomial model using a Dirichlet distribution, parameterised by $\alpha = \alpha_i, \ldots, \alpha_k > 0$, which correspond to the number of times a particular outcome k has been observed. Laplace smoothing is a simple case of this where each parameter is set to one.

To obtain initial parameters we make use of a fixed-point iteration, described in [8]. To update these parameters for each specific case, we simply add the rating counts,

$$\alpha_u = \alpha + \mathbf{n}_u \,. \tag{5}$$

Parameters for our multinomial distribution, can then be obtained by taking the mean of the Dirichlet distribution,

$$E(\mathbf{p}) = \frac{\alpha}{\sum_k \alpha_k}. \tag{6}$$

Adding a Dirichlet prior seriously degraded the performance of the multinomial model. We attempted to improve the model by taking into account variance in the Dirichlet distribution by performing a stochastic expansion about the mean, but this only slightly improved results, so the details are not given here.

6 Gaussian Model

For our next model we decided to look at the differences in user ratings. We model these differences $r_u - r_{u'}$ as being drawn from a Gaussian distribution. Our model is not strictly Gaussian, as we make some simplifications removing constants, and our likelihoods are discrete, rather than continuous. The formula is

$$P(r_{u'} = k' | r_u = k) = \frac{exp(-\tau_{u,u'}/2\,(k - k' - \mu_{u,u'})^2)}{\sum_{k''} exp(-\tau_{u,u'}/2\,(k'' - k' - \mu_{u,u'})^2)} \,, \tag{7}$$

where $\mu_{u,u'}$ is the mean difference between the two user's ratings, and $\tau_{u,u'}$ is the precision of the Gaussian distribution, or σ^{-2} the reciprocal of the variance. We obtain values for the mean and precision through maximum likelihood estimates,

$$\hat{\mu}_{u,u'} = \frac{1}{|I_{u,u'}|} \sum_i (r_{u,i} - r_{u',i}) \,, \tag{8}$$

$$\hat{\sigma}_{u,u'} = \frac{1}{|I_{u,u'}|} \sum_i (r_{u,i} - r_{u',i} - \mu_{u,u'})^2 \,, \tag{9}$$

$$\hat{\tau}_{u,u'} = \frac{1}{\hat{\sigma}_{u,u'}} \,. \tag{10}$$

In cases where there are few ratings, such that $\sigma^2 = 0$, we set the precision to zero.

We can augment this model with a Gaussian-Gamma prior. Mean and variance are treated as unknown, with mean modelled by a Gaussian, and variance by a Gamma distribution. It has the following probability density function,

$$GG(\mu, \tau | \mu, \kappa, a, b) \sim N(\mu | \mu, \kappa\tau^{-1})\Gamma(\tau | a, b) \,. \tag{11}$$

We obtain formulae to calculate its parameters by looking at the marginal distributions of μ and τ,

$$P(\tau) = \Gamma(a,b) \,, \tag{12}$$

$$P(\mu) = T_{2\alpha}(\mu, \frac{a\kappa}{b}) \,, \tag{13}$$

where T is a Student's T distribution[3]. Note that we use the parameterisation of the Gamma distribution where a is the shape parameter, and b is the rate parameter.

We have maximum likelihood estimates for the mean and variance of τ, and using the properties of the gamma distribution, we can derive these estimates for its parameters,

$$a_0 = \frac{\hat{\mu}_\tau^2}{\hat{\sigma}_\tau^2} \,, \tag{14}$$

$$b_0 = \frac{\hat{\mu}_\tau}{\hat{\sigma}_\tau^2} \,. \tag{15}$$

We use a similar procedure using the properties of Student's T distribution to obtain an estimate for κ_0,

$$\kappa_0 = \frac{b_0(a_0 - 1)}{\hat{\sigma}_\mu^2} \,. \tag{16}$$

Finally we set μ_0 to zero as our matrix of differences contains $r_{u,i} - r_{u',i}$ as well as $r_{u',i} - r_{u,i}$, these differences cancel out. Once we have prior values for our parameters, we can perform a Bayesian update using the following equations, which can be found in [3],

$$\mu_n = \frac{\kappa\mu + n\hat{\mu}_{u,u'}}{\kappa_0 + n} \,, \qquad \kappa_n = \kappa_0 + n \,, \qquad a_n = a_0 + \frac{n}{2} \,,$$

$$b_n = b_0 + \frac{1}{2}\sum_{i=1}^{n}(r_u - r_{u'} - \hat{\mu}_{u,u'})^2 + \frac{\kappa_0 n(\hat{\mu}_{u,u'} - \mu_0)^2}{2(\kappa_0 + n)} \,.$$

We obtain parameters for our Gaussian by taking the expected values of the mean and precision from our posterior distribution $GG(\mu, \tau | \mu_n, \kappa_n, a_n, b_n)$,

$$E(\mu) = \mu_n \,, \tag{17}$$

$$E(\tau) = \frac{a_n}{b_n} \,. \tag{18}$$

7 Experiments

To compare the performance of the techniques described in this paper, we implemented several basic recommender system algorithms in Python[1]. We implemented simple recommenders based on using average user ratings and average item ratings to make predictions. We also implemented memory-based CF[2] using Pearson correlation as our similarity measure, and use a fixed neighbourhood size of the 500 most similar users in making predictions, as in our experience this provides the best results. Significance weighting, which weights the effect of users with more items in common, was used with a threshold of 50, as suggested in [5].

The dataset used for our experiments is the MovieLens[2] 100,000 rating dataset, which contains a collection of ratings made by users on films. Ratings are made on a 5-star integer scale. We use this dataset for its popularity which makes comparison with existing methods easier. The dataset is 94 % sparse.

For each experiment we perform 5-fold cross-validation, splitting the dataset randomly by rating, to produce training sets containing 80 % of the ratings, and test sets containing 20 % of the ratings. The same sets are used to test each algorithm. The results of each evaluation are averaged across the five runs.

Our second experiment looks at how these techniques perform under different levels of data sparsity. After splitting each dataset for k-fold cross validation, a varying proportion of the ratings are removed randomly. The results are then tested against the full dataset.

7.1 Evaluation Metrics

We use a small number of commonly used metrics to aid comparison with other techniques, and to give an overall picture of recommender system performance. Recommender system evaluation metrics fall into three main categories: coverage, statistical accuracy, and decision support[5]. In [6] decision support metrics are divided into classification-accuracy metrics, and ranking-accuracy metrics.

We use mean-absolute-error (MAE) to measure statistical accuracy, and the F1 measure to measure classification accuracy. As our ratings are not binary, we transform them by considering ratings greater than or equal to four to be positive, or relevant[5]. We do not use a coverage metric, because we find it does not give useful results. We also do not use a ranking metric, because although the system will return predictions with many different ranks, the ratings system only allows five.

7.2 Results

The results of this experiment are presented in descending order of MAE in Table 1. MAE is presented along with its standard error, and F1 score. Lower values of MAE are better, and higher F1 scores are better. We use a paired t-test at a level of

[1] `http://www.python.org`

[2] `http://www.grouplens.org/`

5 % to test the significance of MAE differences. Methods shown in bold are significantly different from the method below them. Most of the method names are self-explanatory. The variants of the Dirichlet method are corrected and uncorrected, PCC MBCF is memory-based CF using Pearson correlation, with and without significance weighting.

The Gaussian methods perform best on MAE, with the Dirichlet augmented multinomial models performing worst. On F1 score, the Gaussian models come out on top, but in contrast to MAE, the Dirichlet models outperform MBCF. In addition to the Gaussian results shown below which use a Dirichlet model for prior probabilities we tried Gaussian and Gaussian-Gamma priors. We found that these models did not produce significantly different results from the Dirichlet model, so they are not shown here.

Table 1 Results

Method	MAE	SE	F1 Score
Gaussian	0.7054	0.0042	0.6931
Gaussian-Gamma	**0.7059**	0.0042	0.6932
PCC MBCF (Weighted)	**0.7393**	0.0039	0.5613
Laplace	0.7438	0.0043	0.6566
PCC MBCF	**0.7438**	0.0038	0.5481
Item Average	**0.8183**	0.0041	0.3938
User Average	**0.8351**	0.0042	0.2993
Dirichlet (Corrected)	**0.8525**	0.0041	0.5881
Dirichlet	0.8649	0.0045	0.6196

7.3 Sparsity

We looked at a subset of the techniques which performed well in the general experiment for the sparsity experiment. We tested the averages, MBCF with significance weighting; Laplacian, Gaussian, and Gaussian-Gamma naïve Bayes. We tried all variants for the Gaussian models, but found that the Dirichlet model for priors worked best, and so those are the only results reported here for the sake of clarity.

Figure 1 shows MAE against the relative density of the dataset, and Figure 2 shows F1 score against relative density. These graphs show a gradual improvement in performance as more of the original dataset is used. Each techniques maintains its relative performance compared with other techniques for most of the MAE graph. At conditions of extreme sparsity the simple multinomial technique slightly outperforms the others, it is not obvious why this should be so. It is also interesting to note that the Gaussian-Gamma technique, which showed little improvement over the simple Gaussian technique in the general tests, outperforms the Gaussian technique as the data becomes more sparse. It is likely that the prior helps to fill in gaps in its knowledge.

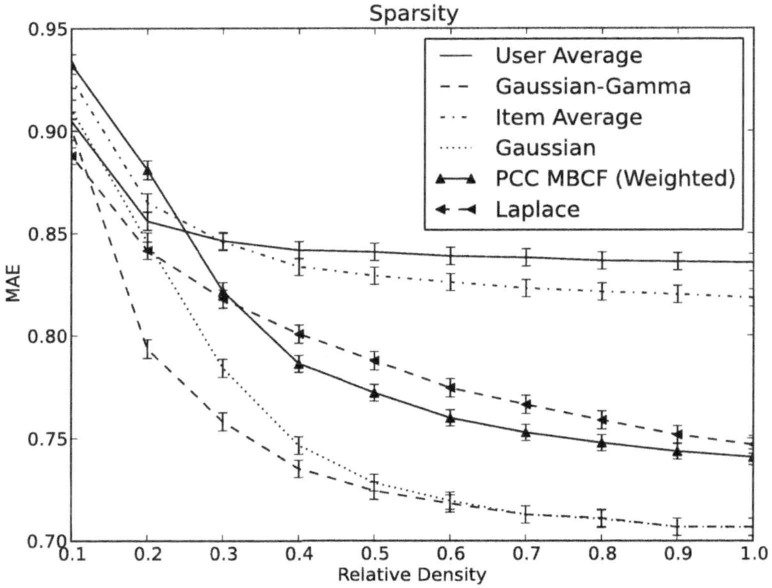

Fig. 1 MAE under conditions of varying sparsity

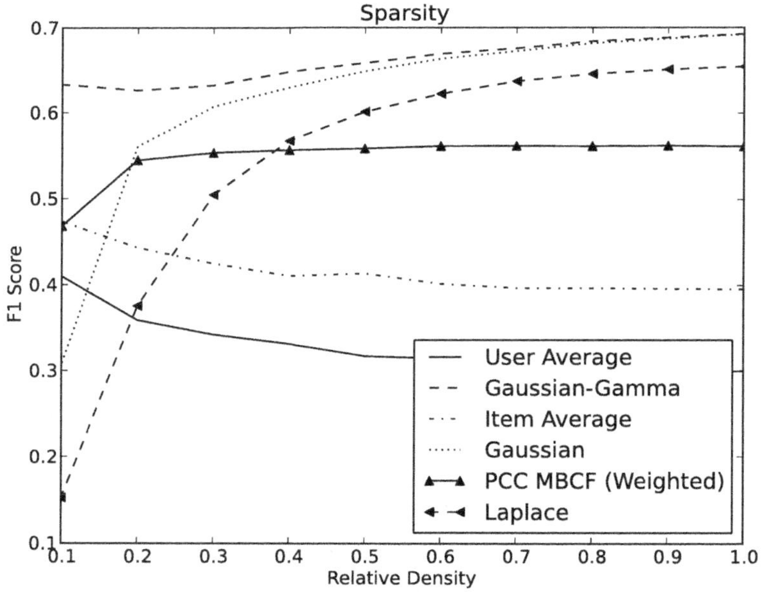

Fig. 2 F1 score under conditions of varying sparsity

The picture is much the same for the F1 measure, although in this case the Gaussian-Gamma technique retains excellent levels of performance even when using the most sparse dataset. It is interesting to note that the performance of the Laplace smoothed algorithm suffers considerably when data is sparse. It is also interesting that increasing data density actually degrades the performance of the average methods.

8 Conclusions

In this paper we have shown that Bayesian recommenders making use of prior knowledge can produce results which are better than those used from memory-based collaborative filtering, and simple probabilistic recommenders not using prior knowledge. We have shown that these techniques maintain good levels of performance even when using sparse data. In particular we find that our Gaussian model produces the best results across most of our tests. Although the Gaussian-Gamma prior was found to perform similarly to the Gaussian model under conditions of relative data density, under sparser conditions it performed better. For our prior probabilities however, we found that the Dirichlet model produced better results.

In the case of the multinomial model, the addition of a prior is found to be harmful to the model. We believe this because we are trying to fit too many parameters given the limited information. In addition the multinomial model makes assumptions about the independence of rating categories which are unlikely to hold true. In cases where the underlying model is a better fit to the data, the addition of a prior appears to help fill in gaps in information leading to better performance in situations where data is sparse.

8.1 Future Work

Our next task is to do a more thorough comparison of our method with existing more complex methods of probabilistic, and sparse recommendation techniques. We also need to apply our technique to a wider range of datasets. As one of our goals is distributed recommendation, we also need to test our method in a distributed situation.

Although we found the Gaussian-Gamma model works well, we still wish to investigate other models. Multivariate Gaussian, or Gaussian mixtures are the next logical steps from this model, but they have more parameters and so might suffer from the same problems as the Dirichlet model. Another approach would be to stay with the Gaussian model, but learn clusters of users, applying different prior knowledge to different groups.

References

1. Adomavicius, G., Sankaranarayanan, R., Sen, S., Tuzhilin, A.: Incorporating contextual information in recommender systems using a multidimensional approach. ACM Trans. Inf. Syst. 23(1), 103–145 (2005), doi: http://doi.acm.org/10.1145/1055709.1055714
2. Breese, J.S., Heckerman, D., Kadie, C.: Empirical analysis of predictive algorithms for collaborative filtering. In: UAI, pp. 43–52. Morgan Kaufmann, San Francisco (1998)

3. De Groot, M.H.: Optimal Statistical Decisions. McGraw-Hill Book Company, New York (1970)
4. Hand, D.J., Yu, K.: Idiot's bayes—not so stupid after all? International Statistical Review 69(3), 385–398 (2001)
5. Herlocker, J.L., Konstan, J.A., Borchers, A., Riedl, J.: An algorithmic framework for performing collaborative filtering. In: SIGIR 1999: Proceedings of the 22nd Annual International ACM SIGIR Conference on Research and Development in Information Retrieval, pp. 230–237. ACM, New York (1999)
6. Herlocker, J.L., Konstan, J.A., Terveen, L.G., Riedl, J.T.: Evaluating collaborative filtering recommender systems. ACM Trans. Inf. Syst. 22(1), 5–53 (2004)
7. Maltz, D., Ehrlich, K.: Pointing the way: active collaborative filtering. In: CHI 1995: Proceedings of the SIGCHI Conference on Human Factors in Computing Systems, pp. 202–209. ACM Press/Addison-Wesley Publishing Co., New York, NY, USA (1995), doi: http://doi.acm.org/10.1145/223904.223930
8. Minka, T.P.: Estimating a dirichlet distribution. Tech. rep., Microsoft (2003)
9. Miyahara, K., Pazzani, M.J.: Collaborative filtering with the simple bayesian classifier. In: Proceedings of the 6th Pacific Rim International Conference on Artificial Intelligence, pp. 679–689 (2000)
10. Resnick, P., Iakovou, N., Sushak, M., Bergstrom, P., Riedl, J.: Grouplens: An open architecture for collaborative filtering of netnews. In: Proc. Computer Supported Cooperative Work Conf. (1994)
11. Sarwar, B., Karypis, G., Konstan, J., Reidl, J.: Item-based collaborative filtering recommendation algorithms. In: WWW 2001: Proceedings of the 10th International Conference on World Wide Web, pp. 285–295. ACM, New York (2001)
12. Tveit, A.: Peer-to-peer based recommendations for mobile commerce. In: WMC 2001: Proceedings of the 1st International Workshop on Mobile Commerce, pp. 26–29. ACM, New York (2001), http://doi.acm.org/10.1145/381461.381466
13. Wang, J., Robertson, S., Vries, A.P., Reinders, M.J.: Probabilistic relevance ranking for collaborative filtering. Inf. Retr. 11(6), 477–497 (2008)

Experiences of Knowledge Visualization in Semantic Web Applications

Nadia Catenazzi and Lorenzo Sommaruga

Abstract. There is an increasing need for usable tools to support knowledge elicitation, formalization and management. As an answer to this need, this paper describes fully integrated semantic web framework experiences, where users can represent and manage their data in a visual way, without the need of semantic web experts as intermediaries. These frameworks typically incorporate an ontology editor, a resource editor, reasoning capabilities and intuitive interaction and visualization facilities. The use of effective visualization techniques to graphically represent ontologies is investigated and the EasyOnto prototype is presented. In addition, different semantic web frameworks have been implemented as a result of research projects in different domains. In particular this paper presents the IRCS framework, an intelligent software to semantically index, search, and navigate the documentation used in water management plants, and the AWI environment, a collaborative environment aiming to collect and share knowledge of user communities, within the context of a digital factory.

1 Introduction

Nowadays, there is an increasing need for flexible and intuitive tools to support knowledge elicitation, formalization, management and sharing, in a number of different fields. The usefulness of this sort of tools emerges for instance in the context of business process management and digital factories (Sommaruga et al. 2010). Other domains where a similar need has come out are mass customization and water management treatment plants. In a regional context, such as in the southern Switzerland area, where SMEs are the typical industrial reality, the SUPSI academic institution has received various technology transfer requests

Nadia Catenazzi . Lorenzo Sommaruga
Semantic and Multimedia Systems Lab. (LSMS), DTI Dept. (SUPSI-DTI)
University of Applied Sciences of Southern Switzerland, CH-6928 Manno
e-mail: {nadia.catenazzi,lorenzo.sommaruga}@supsi.ch

E. Mugellini et al. (Eds.): Advances in Intelligent Web Mastering - 3, AISC 86, pp. 49–59.
springerlink.com

from the real world and has undertaken a number of applied research projects, where knowledge management is the primary focus (KTI project nr. 9402.1 PFES-ES, KTI project nr. 9070.2 PFES-ES http://www.difac.ch).

Semantic web technologies can offer an answer to this need; they provide formalisms to describe domain entities and relations (i.e. ontologies), creating a metadata level for interoperability among heterogeneous data, and offer mechanisms to infer new knowledge starting from the explicitly declared one. Existing semantic web tools are powerful systems to represent and manage knowledge, but they are often oriented to ontologists and not to domain experts. From a survey reported in (Denny 2004), it already emerged that there were various enhancements required by users in future ontology editors. The improvements most often mentioned by respondents were "a higher-level abstraction of ontology language constructs to allow more intuitive and more powerful knowledge modeling expressions", easy understanding of the ontology improving visual/spatial navigation, and more options for using reasoning facilities.

Considering the inadequacy of existing systems to answer the need of real world "customers" in our applied research projects, the interest is for developing fully integrated semantic web frameworks where users can represent and manage their data in a user-friendly way, without the need of semantic web experts as intermediaries. Although the ontology is an essential part of the system, the users' primary focus is on their world of resources that often represents heterogeneous interrelated data. Domain experts who have competences to create the conceptualization of the domain, i.e. the ontology, do not usually speak the RDF and OWL terminology. Therefore a higher abstraction level of the ontology constructs is required, by means of a visual interactive representation that is a simplification of the world, and allow ontology management to happen in a transparent way. In order to meet this requirement, two investigation directions have been followed:

- on one hand, the use of effective visualization techniques to graphically represent knowledge, i.e. ontologies and the interrelated resources;
- on the other hand, the development of integrated development environment for semantic web applications, which mainly facilitate ontology and resource editing.

The rest of this paper summarizes the main outcomes of our experiences.

2 Knowledge Visualization in the Semantic Web Context

Visualization is an effective way to communicate abstract data and information through the use of interactive visual interfaces. Knowledge visualization tools have a great potential as they allow knowledge, usually coded in a rigorous but not intuitive formalism, to be visually presented and therefore easily perceived. Visualization tools can be considered mechanisms for knowledge valorization.

Within the semantic web context, knowledge visualization can cover both resource description aspects and ontological aspects. Geroimenko (2006) describes different approaches and techniques to visualize semantic data and metadata, topic maps, ontologies, etc.

The strategic role of *ontology visualization* tools in the process of ontology engineering has been widely recognized in literature (Katifori 2007, Lanzenberger 2010, Catenazzi et al. 2009). Ontology visualization tools help to highlight the class structure and the relations among classes. Katifori (2007) provides a detailed analysis of the existing visualization methods, identifying the following categories: indented list (i.e. a class browser following the file system explorer-tree view), node-link (i.e. graph) and tree, zoomable, focus + context, space-filling, and 3D Information Landscapes.

With respect to *resource visualization*, a distinction is needed between resources that are individuals defined as class instances of an ontology, and RDF resources that are independent from the ontology. In the first case it is possible to show the instances connected to their membership classes using ontology visualization methods; an example of this approach is adopted in the Jambalaya Protégé plugin (http://www.thechiselgroup.org/jambalaya). When RDF resources are not connected to an ontology, a typical visualization is the graph view: instances are represented as graph nodes and relations as arcs. This method can also be applied when resources are defined as class instances. In this case, different visual clues, such as color, shape o icon, can be used to distinguish individuals belonging to different classes. When the number of nodes and relations is very high, the graph visualization of RDF resources can produce clumsy and unreadable views, generating information overload. In this case filtering mechanisms can represent a useful aid to support information selection. Another effective solution can be to reduce the complexity giving a partial representation of the resource world; a possible approach is to visualize the node of interest at the centre of the graph surrounded by the directly connected nodes, as in the Star Resource Navigator (Catenazzi and Sommaruga 2005b) or surrounded by those nodes that have a specific "distance" from the central node. This approach has been proven to be beneficial within a learning context in visualizing interconnected resources about people, competences and courses (Catenazzi and Sommaruga 2005a).

In our works many techniques have been tested for both ontology and resource visualization. The EasyOnto project, described in the next section, is an experimental work implementing various ontology visualization methods.

2.1 EasyOnto

EasyOnto project is a web application designed for ontology visualization, which has been carried out at LSMS SUPSI DTI in partial fulfillment for a bachelor thesis (Ruggeri 2009). In this work, navigation tools and different visualizations methods are exploited to allow the user to get a fully interactive experience.

EasyOnto combines a textual and a graphical representation of OWL ontologies (see figure 1). In the textual representation, shown at the bottom of the window and always visible, classes are represented in a class tree, using the intended list method. By selecting a class, its data and object properties are shown. In order to simplify the user interaction, technical terms have been replaced by more intuitive

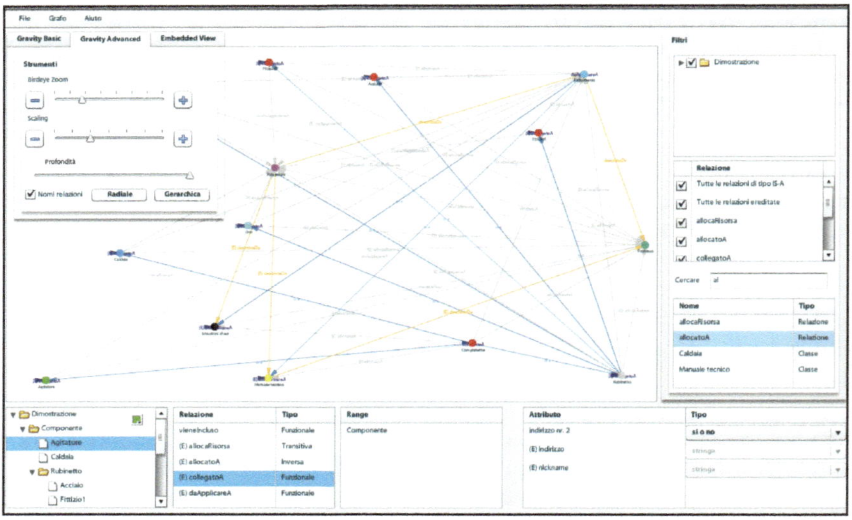

Fig. 1 EasyOnto: a graphical representation of the ontology using a graph

ones (for instance relations is used instead of ObjectProperty) and class and property names are visualized without their namespace.

The graphical representation of the ontology takes the most part of the window. Colors are used to visually distinguish classes and relationships, and some configuration controls are available to change visualization parameters such as scaling, zooming, node level depth setting and view mode (hierarchical/radial).

The EasyOnto system provides a number of facilities to differently present an ontology structure. There are basically two alternatives to represent the ontology: one visualization based on the graph method, and a more advanced one using a zoomable technique. In the first case the traditional gravity based visualization of graphs is used, provided by libraries such as SpringGraph (http://mark-shepherd.com/SpringGraph) or BirdEye (http://code.google.com/p/birdeye).

The second visualization type is based on the OWLeasyViz visualization strategy (Catenazzi et al. 2009), which integrates the zoomable and graph ontology visualization methods (see figure 2). This approach offers a more clear view of the ontology because it separates hierarchical relations from role relations, i.e. non-hierarchical relations. Role relations are represented as arcs connecting the source and destination nodes, as in the graph method. Hierarchies are represented as nested sets. Child nodes are visualized inside their parents, with smaller size. When the user clicks on a child node, the node is expanded and its content is made visible, using the zoomable technique. Existing instances are shown at the bottom level of the hierarchy. This ontology visualization model exploits the visualization strategies similar to those used in Jambalaya, a Protégé plug-in specifically developed for ontology visualization (http://www.thechiselgroup.org/jambalaya).

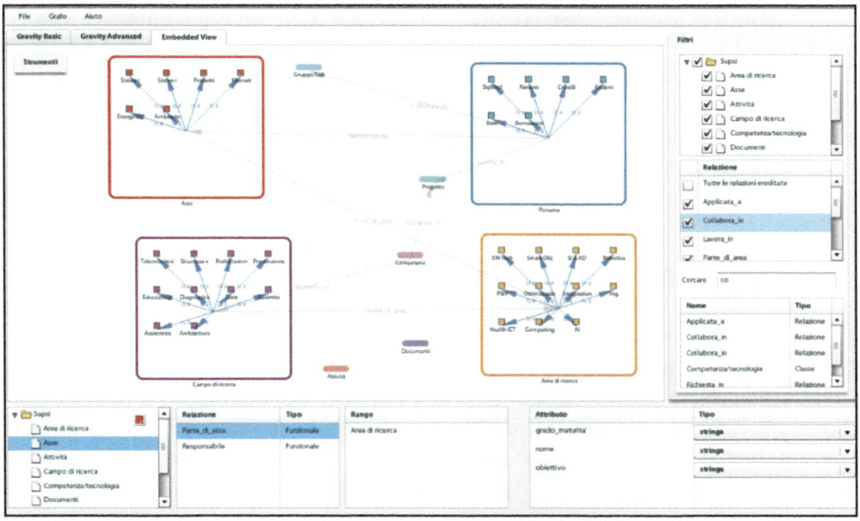

Fig. 2 EasyOnto: The OWLeasyViz graphical representation integrating zoomable and graph visualization methods

In order to simplify and reduce the visual complexity and facilitate ontology access, navigation, and visualization, a filtering functionality has been integrated. Only the selected classes and properties are shown both in the textual and graphical representation of the ontology, facilitating direct access to resources.

This project was mainly an academic work, aimed to demonstrate the practical application of some of the ontology visualization methods presented in literature, by using and adapting existing graphic libraries. Some of its ideas and graphical solutions have been adopted in applied research projects, as described in the next section.

3 Integrated Environments for Knowledge Management and Visualization

This section describes two systems implemented as the result of research projects in different domains: the IRCS Water, an intelligent software to semantically index, search, and navigate documentation used in water management plants, and AWI, a collaborative environment aiming to collect and share knowledge of user communities.

Some of their ideas and principles are taken from the Semantic DB (Mantegazzini 2008), a preliminary work developed as an academic research prototype. The Semantic DB is a web application framework to create simple semantic web applications, where the user can simultaneously operate both on the model of the world and on its individuals. The Semantic DB integrates an ontology editor, an RDF resource editor and navigator, a text-based inference rule editor, a reasoner to make inferences on the basis of the defined rules, a search engine to find resources, and a path search engine to identify paths among any pair of resources. Although the ontology editor does not provide full support to the OWL specification, simple ontologies and semantic world of resources can be defined, showing the applicability of semantic web technologies to non-expert people and their potential benefits. This system was mainly used as a didactic tool and as a show-case to demonstrate the potential applicability of semantic web technologies.

3.1 IRCS Framework

IRCS Water is a Swiss Government funded applied research project in the domain of waste water management (CTI project nr. 9402.1 PFES-ES). In the context of water management, an important issue is to manage the huge amount of documentation and heterogeneous data, coming from different sources and available in different formats. Examples of documents are plant and component manuals, maintenance procedures, meeting reports, synchronous and asynchronous communications, etc. Some of these resources are available in digital form, other exist only on paper, others are part of the workers' knowledge. Although these resources represent an important knowledge source, their heterogeneity, the lack of structure, the lack of metadata and explicit relations, make difficult and slow to retrieve specific resources. The project goal was to develop an intelligent software to semantically index, search, and navigate the documentation used in water management plants. Semantic web technologies are adequate to satisfy this kind of need, allowing reference models (ontologies) to be defined for data. In order to reach the project objective, a generic environment to create semantic web applications, called *IRCS Framework*, was developed and successively applied in the context of water management plants. The resulting application is called *IRCS Water*. The IRCS Framework consists of different modules:

- *ontology module*: where the models of resources are defined (OWL ontology creation and editing);
- resources module: where resources are acquired and modified according to the models;
- *inference module*: where implicit knowledge is captured and codified in form of inference rules, and applied to deduce new knowledge;
- *administration module*: where projects, models, users are managed.

The ontology module (see figure 3), similarly to other projects, provides functionalities to create and modify classes, properties and relations; it also offers a graphical visualization of the ontology using the graph method, and a filtering mechanism, which allows information of interest to be selected.

The resources module (see figure 4) provides functionalities to create and modify resources, their properties and relations; it also offers a graphical visualization of the resource world, searching and filtering mechanisms, which allow information of interest to be searched or selected for visualization. As in the ontology module, the graphical visualization of resources is based on the graph method.

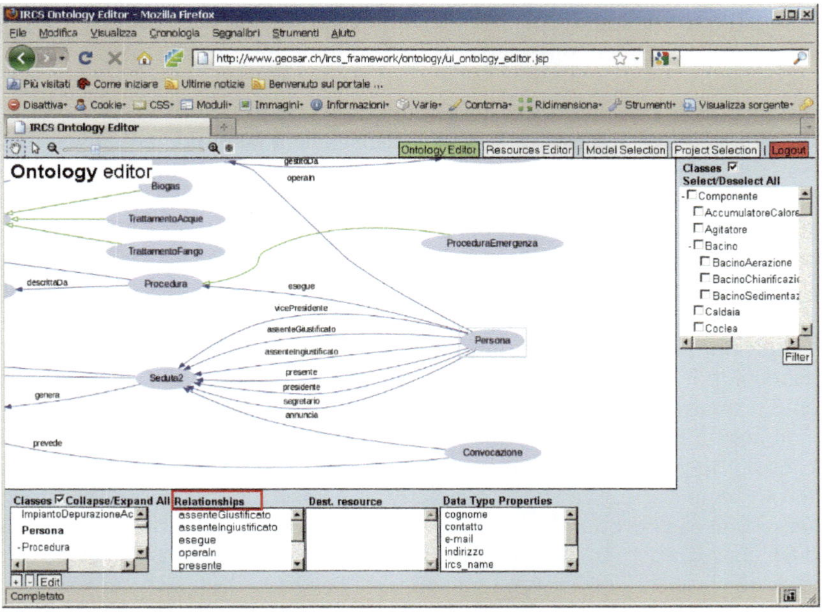

Fig. 3 Ontology Editor in the IRCS framework

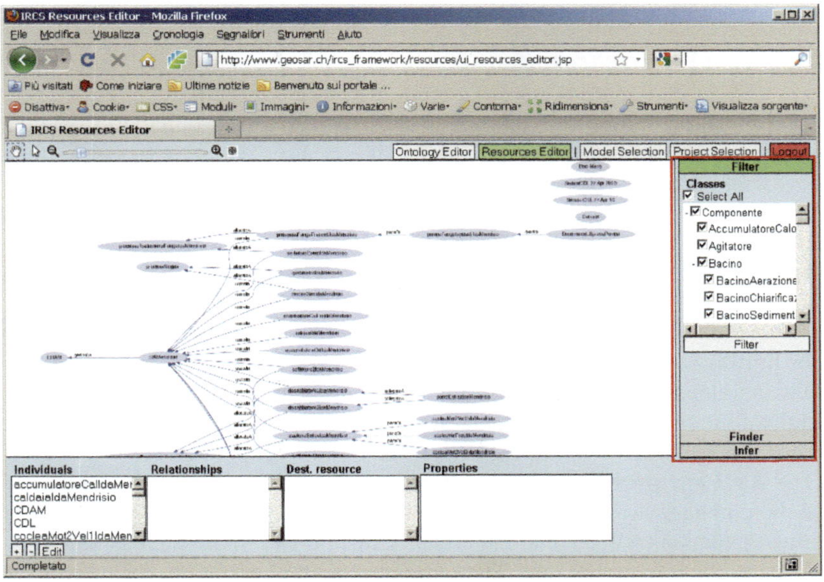

Fig. 4 Resource Editor in the IRCS framework: filtering mechanism

The inference module is accessible from the resource editor through the *infer* button. Two basic functions are supported: rule definition and rule selection and application.

An empirical evaluation was conducted to assess the framework usability and effectiveness in the context of the water management plant domain, involving real users. The ontology editor received a positive evaluation. Domain experts, who are the primary users of this module, have recognized its potential and effectiveness in modeling the domain resources, and have found the adopted visualization strategy easy to understand. Considering the application domain of the IRCS water project, the solution adopted for ontology editing and visualization has resulted to be effective: the number of classes to represent the water plant domain is limited and the filtering mechanism is useful to select only information of interest. The resource editor has received a different evaluation by real users, who are mainly people responsible for the documentation (e.g. secretaries), who needed to insert or automatically import huge amount of documentation related to water plants. Because of the large amount of involved resources, in this case the editor has resulted to be not effective and not directly usable for this purpose. The graphical visualization can become clumsy and unreadable, making difficult to insert data.

The consequence is that ad-hoc visualizations are necessary. An additional module for inserting or importing documents has been designed and is under development. This software add-on provides a specific graphical user interface and exploits the core functionalities of the IRCS framework, relying on its three-tier SOA based architecture.

3.2 The AWI Environment

AWI (Baldassari 2009, Sommaruga et al. 2010), is a web collaborative environment aiming to collect and share knowledge of user communities. The peculiar aspect with respect to the IRCS framework is that it integrates in a single working environment both the tools required to work with a semantic world (semantic web tools) and the tools needed to communicate and share knowledge (social web tools). Semantic web tools basically include an ontology editor and a resource editor. The ontology editor provides simple facilities for creating, deleting and modifying classes, properties and relations, as well as OWL ontology importing/exporting capabilities. The resource editor provides tools to create, modify and delete individuals, properties, and relations on the basis of a specified ontology. Social web tools include wiki, social tagging, community tagging, shared calendar and content ranking.

The AWI environment has been incrementally developed: in the original prototype the emphasis was mainly on social web aspects (see SmartBricks below), the following version, AWI2, also integrated semantic web tools such as ontology and resource editing, querying, etc. The current version, AWI3, introduces two important improvements: on one hand, it provides more usable interaction facilities and visual interactive representations of the semantic world; on the other hand, it provides advanced reasoning capabilities on it. Both the ontology and the resources are represented as a graph. Figure 5 shows an example of graph visualization in the

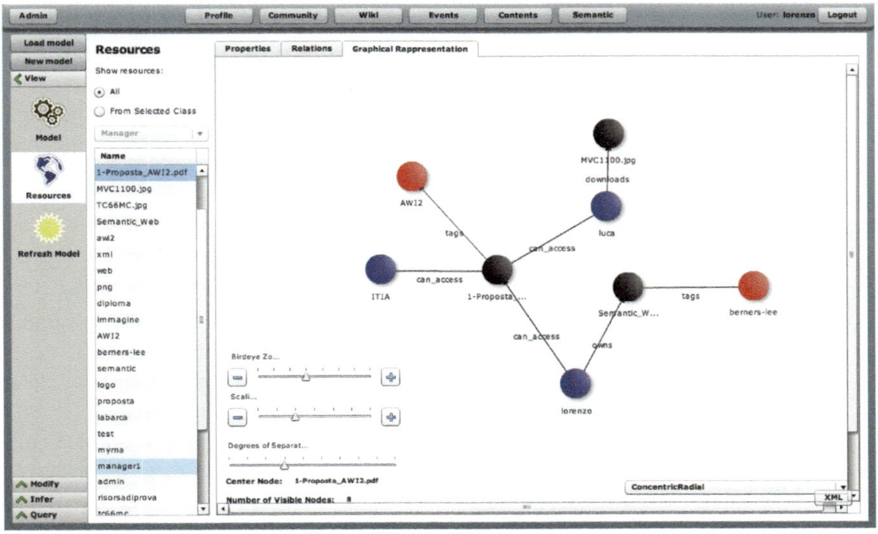

Fig. 5 AWI3 - graphical semantic world representation

context of document management. The semantic world is represented as a graph of interconnected resources belonging to different classes. A specific user assigned color identifies individuals of a given class. For instance, red resources are tags, that are connected to black resources, i.e. documents. Zooming, scaling and other functionalities are also provided to modify/configure the graph visualization.

Inference rules can be defined using a graphical wizard or directly using the SWRL formalism. Once rules are defined, users may select which rules to apply and activate a reasoner to infer new knowledge. The application of an inference rule may produce changes in the world of resources; for instance new relations may appear among resources that were not previously interconnected. These changes are visualized in the graph, that represent changes in a different color.

As shown in the examples, the AWI environment was used to create ontologies in various domains such as family and business process management. The SmartBricks prototype (Flores and Sommaruga, 2009) is the result of the AWI application in the field of manufacturing and business process management. It was developed within the IMS-KTI DiFac project, whose purpose was to develop intelligent tools for Business Processes Management (BPM) to be used by a multinational company.

The system design was based on the analysis of the company needs, which leads to the adaptation of the generic tools to meet the enterprise specific requirements. The SmartBricks prototype was tested by more than 50 business process managers. Results produced by the preliminary test were encouraging and demonstrated the potential of such a system in a real context.

4 Conclusions

In the context of knowledge representation, managing and visualization, this paper has described the IRCS Framework and the AWI environment. The followed approach was to develop generic environments based on semantic web technologies and to apply them to the specific domains (water management plants, business process management, etc.). The generic nature offers a great potential and applicability and, at the same time, introduces usability problems to some user categories. For these users, ad-hoc solutions are surely more effective.

The two frameworks provide many common functionalities:

- an *ontology editor*, where classes, data properties and object properties can be created and modified;
- a *resource editor*, where individuals of the ontology classes can be created, modified and searched (full-text and SPARQL query are supported);
- an *inference mechanism* which allow users to define and apply inference rules. AWI provides a wizard to define rule, while IRCS requires rules to be written in Jena rule language;
- *visualization facilities* to show both the ontology and the instances using the graph method. Filtering, zooming mechanisms are available in both systems to improve the visualization. AWI provides more advanced visualization features (e.g. degree of separation).

Although the systems implement the basic constructs of the OWL specification, they are enough expressive to describe non-complex domains. There are functionalities which are peculiar of the AWI system: they basically include social web tools such as wiki, social tagging, and shared calendar.

The key aspect of these systems is to provide intuitive and easy to use environments which can be used by domain experts, who are not semantic web experts. A crucial role in reaching this objective is played by visualization tools. Our experiences in different domains confirm that there is not a single method that can be considered the best in any situation. For instance, the graph method to visualize ontologies, supported by filtering mechanism, has demonstrated to be effective in the application domains, where the number of classes is generally limited. The same method applied to resources can work well when the resource world is relatively small (e.g. the family domain used in the AWI environment), but can become unreadable with a large resource set.

References

Baldassari, A.: Sviluppo Applicazione Web Intelligente, AWI2, Comp. Sc. Eng. Bachelor Thesis, SUPSI DTI (2009)

Catenazzi, N., Sommaruga, L.: A Semantic Web Resource Navigator for Competence Based Learning. In: Proceedings of International Conference on Methods and Technologies for Learning, ICMTL 2005, Palermo Italy, March 9-11 (2005a)

Catenazzi, N., Sommaruga, L.: Practical experiences towards generic resource navigation and visualization. In: Proceedings of SWAP 2005, the 2nd Italian Semantic Web Workshop. CEUR Workshop Proceedings, Trento, Italy, December 14-16 (2005b) ISSN 1613-0073, http://ceur-ws.org/Vol-166/58.pdf

Catenazzi, N., Sommaruga, L., Mazza, R.: User-friendly ontology editing and visualization tools: the OWLeasyViz approach. In: 13th International Conference Information Visualisation, Barcelona (July 2009)

Denny, M.: Ontology Tools Survey (2004), http://www.xml.com/pub/a/2004/07/14/onto.html

Flores, M., Sommaruga, L.: SMARTBRICKS: Developing an Intelligent Web tool for Business Process Management. In: 15th International Conference on Concurrent Enterprising, Leiden - The Nederlands, June 22-24 (2009)

Gašević, D., Djurić, D., Devedžić, V.: Ontologies in "Model Driven Architecture and Ontology Development", ch. 2. Springer, Heidelberg (2006)

Geroimenko, V., Chen, C. (eds.): Visualizing the Semantic Web, XML-based Internet and Information Visualization. Springer, Heidelberg (2006)

Katifori, A., Halatsis, C., Lepouras, G., Vassilakis, C., Gainnopoulou, E.: Ontology visualization methods - a survey. ACM Computing Surveys (CSUR) 39(4) (2007)

Lanzenberger, M., Sampson, J., Rester, M.: Ontology Visualization: Tools and Techniques for Visual Representation of Semi-Structured Meta-Data. Journal of Universal Computer Science 16(7), 1036–1054 (2010)

Mantegazzini, R.: Semantic DB, Advanced Computer Science Master Thesis, SUPSI DTI (2008)

Ruggeri, A.: EasyOnto Visualizzazione ontologie, Comp. Sc. Eng. Bachelor Thesis, SUPSI DTI (2009)

Sommaruga, L., Catenazzi, N., Canetta, L.: The Intelligent web: tools for knowledge management and sharing. In: Proceedings of 16th International Conference on Concurrent Enterprising "Collaborative Environments for Sustainable Innovation" (ICE 2010), Lugano – Switzerland, June 21-23 (2010)

"Tagsonomy": Easy Access to Web Sites through a Combination of Taxonomy and Folksonomy

Lorenzo Sommaruga, Petra Rota, and Nadia Catenazzi

Abstract. This paper analyzes possible solutions and mechanisms to facilitate information retrieval in a corporate web site. It presents advantages and problems of taxonomies and folksonomies, and proposes a hybrid approach which combines their benefits to adaptively improve access to a web site. This approach has been tested in a real scenario by developing the Easy Access system. This practical experience introduces the "tagsonomy" mechanism, i.e. the controlled combination of a top-down classification defined by the web site content manager, and a bottom-up classification defined by users. The peculiar feature is that the folksonomy is automatically generated on the basis of the user interaction with the system, and not as the result of an explicit tagging process.

Keywords: folksonomy, taxonomy, tagsonomy, social classification, eGovernment, easy access, information retrieval.

Lorenzo Sommaruga · Nadia Catenazzi
University of Applied Sciences and Arts of Southern Switzerland (SUPSI),
Lab. for Semantic and Multimedia Systems (LSMS),
DTI-ISIN, Via Cantonale - Galleria 2
CH-6928 Manno
e-mail: `{lorenzo.sommaruga,nadia.catenazzi}@supsi.ch`

Petra Rota
Repubblica e Cantone Ticino
Bellinzona, Ufficio della comunicazione elettronica,
Palazzo delle Orsoline
CH-6500 Bellinzona
e-mail: `petra.rota@ti.ch`

E. Mugellini et al. (Eds.): Advances in Intelligent Web Mastering - 3, AISC 86, pp. 61–71.
springerlink.com

1 Introduction

One of the most important objective of a web site developer is to organize contents in such a way that users are able to easily retrieve information of interest. A frequently used analogy to explain how human users search for information in the web is to compare them to wild beasts in the jungle looking for food (the so called informavores) [1]. Users look for maximum benefit with minimum effort. It is important that content appears as a nutritious meal (good appealing content) and that it is easy to find. A crucial aspect to be considered in order to help users to retrieve information they seek is the information architecture, i.e. how the information space is structured [2].

Structuring a web site is generally a complex task which involves several issues such as usability, visual design, user experience, orientation, navigation, accessibility, etc.

The site structure has to match user expectations, the navigational structure has to reflect the end-user view of the site and not the developer view. The choice of link and menu item labels should be consistent with what users will find at the destination. Navigational menus are usually hierarchically organized. The menu categories and sub-categories should be clearly organized to facilitate user access to information, avoiding information overload and filtering the most important contents. As a complement to navigation mechanisms, an internal search engine is often provided. This tool should be carefully designed. Internal search engines that do not work properly represent a greater source of frustration than not having one at all, and may give the user the impression that a specific content is not available on the site even if it is. The origin of this problem is that the search keywords introduced by the user do not correspond to the keywords used to index contents. In conclusion, the difficulty for a user to retrieve information of interest, using both navigation and search mechanisms, often depends to the fact that "user terminology" (and mental model) often does not match the "site" one.

This paper analyzes possible solutions and mechanisms to facilitate information retrieval in a web site by adapting the content developer point of view according to final users' interests. In particular, it presents advantages and problems related to the use of taxonomies, i.e. predefined classifications, and folksonomies, i.e. user generated classifications. In addition, it proposes the hybrid approach, that integrates collaborative tagging (folksonomy) and top-down classifications (taxonomy), as a solution to most of the limitations of the two methods. Finally, it describes the Easy Access system, a practical experience of integration of the two approaches, applied to a real organization web portal.

2 Web Access through Taxonomies and Folksonomies

A taxonomy is a particular classification usually arranged in a hierarchical structure. The term, originally referred to the classification of organisms, was later extended to consider classifications in any domain. In principle, the same "term" can be classified in many different ways according to the expected use and the person who organizes it. Taxonomies are designed by specialized staff; they are accurate, precise and, at the same time, rigid schemes which may produce ontological definitions.

In the web context, taxonomies were historically used in form of web directories to classify web sites in categories and sub-categories [3]. Typical examples of general directories are Google and Yahoo! Directory. Within web sites, the taxonomic approach is traditionally used to organize information in hierarchical menus that should guarantee easy access to the corresponding web pages.

Taxonomies are usually defined by content experts, but may also be automatically generated. Some Content Management Systems provide tools that can be used to classify the site contents. An example is Drupal with its Taxonomy module useful to organize and catalogue the contents of a web site [4].

Nowadays new approaches are emerging to facilitate information retrieval based on tagging systems. Thanks to the increasing use of tags inserted by web users, folksonomies became popular starting from 2004. Typical applications where this approach is followed are Delicious and Flicker. Folksonomies are classifications generated by users in a collaborative way, to annotate and classify a specific content (an image, a video, a text, etc.). Collaborative tagging reflects user mental models, and over time brings out the collective vision of information, because people can use their own vocabulary. Folksonomies are therefore flat, uncontrolled, bottom-up classification systems that emerge from social tagging. Tags are defined by both content creators and consumers, rather than by experts.

Both taxonomies and folksonomies present positive and negative aspects [5,6]. Taxonomies have a number of benefits: they are precise and accurate classification schemes, as already mentioned, and work well with a known-item seeking strategy i.e. when users know what they are looking for. Their limitations mainly include rigidity, closure (difficult to insert a new category), and centralization (designed by few experts). In addition, they do not suit exploratory-seeking strategies. Folksonomies have a number of strengths: they reflect users' mental model, they incorporate multiple perspectives, they match users' language and need, and they are suited for serendipitous discovery (exploratory-seeking strategies). Their main drawbacks include: lack of precision, uncontrolled vocabulary (ambiguity, inconsistent vocabulary, no synonym and polysemy control), and flat structure (no hierarchy).

3 Combining the Taxonomy and Folksonomy Approaches

Taking into account the potential of taxonomies and folksonomies, an innovative approach combining the two methods is here explored. Some main categories are top-down defined (taxonomy) and successively bottom-up improved through collaborative tagging. The hybrid method merges the freedom of social tagging with some control coming from the top-down classification. This approach is considered a very promising solution to improve web access, and is currently implemented in real contexts/systems.

An example of integration of top-down and bottom-up classification is provided by the "Tags for citizens" project, applied to the Turin municipality website (http://www.comune.torino.it). The taxonomy is based on the British standard IPSV (Integrated Public Sector Vocabulary), while the bottom-up classification comes from a social tagging system [5]. The folksonomy layer has two purposes there:

- to allow users to label pages of interest through tags and save them in a reserved area
- to allow users to use tags created by others as a complementary browsing tool to the taxonomy and search engine.

In this system the folksonomy layer is integrated into the existing taxonomy: when users insert a new tag, they are asked to link it to the taxonomy categories creating a link between the two layers.

Another experience of integration of bottom-up and top-down classification is FaceTag [7] (http://www.facetag.org/), "a working prototype of a semantic collaborative tagging tool conceived for bookmarking information architecture resources". The combination of the flat tag space created by users and a richer faceted classification scheme contributes to disambiguate and contextualize tags, may help to solve the linguistic issue of folksonomies (i.e. polysemy, homonymy and base level variations, etc.), and to improve the system information architecture.

Another site, where the need for a hybrid approach has been recognized, is the BBC web site [8]. The proposed solution, known as "Metadata Threshold", allows user tags to be added to control vocabularies, when enough content is tagged with that term.

4 The Easy Access (EA) Project

As already mentioned, the traditional searching and navigation mechanisms are often not enough to help users find information they are looking for. The EasyAccess Project [9] general goal is to provide a solution to the problem of finding information of interest in a web site. As the systems described above, this project uses a combination of a top-down classification defined by the web site content manager, and a bottom-up classification defined by users. We coined the new term "tagsonomy" (tag + sonomy) to refer to the above combination, making it adaptive in context and time. In fact, the peculiar feature, with respect to other systems that use a similar approach, is that the folksonomy is automatically generated on the basis of the user interaction with the system, and not as the result of an explicit tagging process (see figure 1).

More specifically, the taxonomy is a predefined classification of subjects, created by the content manager, organized as a hierarchy of categories and subcategories, and used in the navigation menu. Each entry of this hierarchy is a-priori associated to a page with a specific URI. From this taxonomy, a set of predefined terms are extracted to become tags of the tagsonomy; they usually are category and sub-categories names.

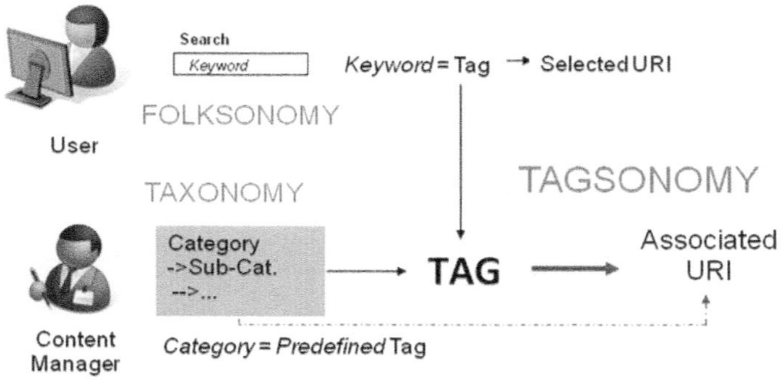

Fig. 1 Tagsonomy: integration of a folksonomy and a taxonomy

The folksonomy is incrementally created on the basis of the keywords inserted by the user during the search process offered by the web site. The search keyword becomes a tag associated to the page the user selects in the search result list. In other words, the user associates a link (URI) to a keyword by selecting the most promising item in the list of search results. All user activities are tracked in a log file and used to create the tagsonomy. Misleading associations assigned by mistake by users in this step can be solved later by a control mechanism in the administration console, described below.

If the user tag corresponds to an existing predefined tag, and the associated URI is different, the new URI (i.e. link) is added to the tag in addition to the existing ones. The final association of a tag to a specific page URI depends on the frequency of use. If many users associate an URI which is different from the one defined by the content manager, the user URI will become "stronger" than the predefined one.

The resulting tagsonomy will consist of a number of tags, some coming from the taxonomy (predefined tags), some generated from the folksonomy (searched keywords), and some others derived from predefined tags altered through the user search. Initially, the tagsonomy will mainly contain taxonomy terms, and, afterwards, it will evolve towards a controlled folksonomy on the basis of the user interaction.

The tagsonomy is visualized as a Tag Cloud where tags have a specific colour and size (see figure 2). Different colours are associated to the different tags identifying the context according to the destination link.

Fig. 2 An example of the visualization of an EA Tag cloud derived from the tagsonomy

The size depends on a number of parameters that are implicitly defined by the user and explicitly by the content manager. The user factors mainly include: the search frequency of the tag, the number of times user associated the specific URI to the tag by selecting it in the search result list, the frequency of the associated URI activation (taking into account all possible link activation modes e.g. direct activation, activation through a menu), and the time elapsed since the last click. Content manager factors mainly include a priority (i.e. importance of the tag), and a temporal factor (i.e. relevance of the tag in a specific time period). It is possible to increase or decrease the relative importance of these parameters by giving them a different weight in a formula which determines the tag cloud text size.

The resulting tagsonomy is therefore an adaptive flexible structure that can be modified giving more priority to the user factors (i.e. the folksonomy) or to the content manager factors (i.e. the taxonomy). In addition, tags can be filtered using a list of badwords.

In order to manage the tagsonomy, an administration console was designed and implemented (see figure 3). This tool is useful to manage tags and links, to configure the layout of the tag cloud (size, color), to define the weight of the different parameters in the formula, to define the list of badwords, and other functionalities.

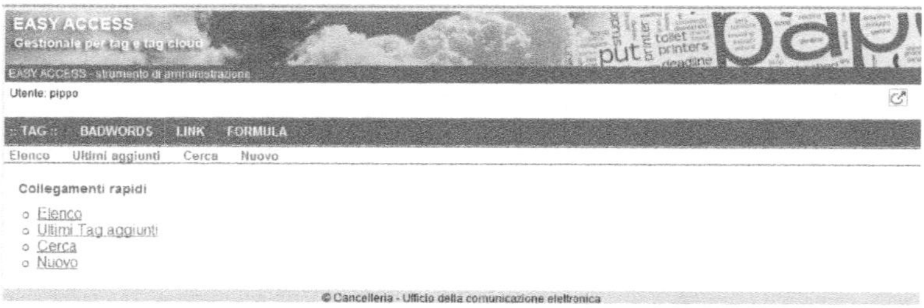

Fig. 3 The EA administration console

4.1 Test Case: Applying the EA Tagsonomy to a Web Site

This project was born as a Bachelor degree Thesis at SUPSI to meet a specific need of a corporate web site: the "Repubblica e Cantone Ticino" web portal (http://www3.ti.ch/), containing general information by the local administration. These are contents of public interest, mainly oriented to citizens and to local administrators. It is worth noting that the tagsonomy approach could be applied to other similar sites where there is a large and complex amount of information, an heterogeneous target user, and high access rate (thousands of searches/day).

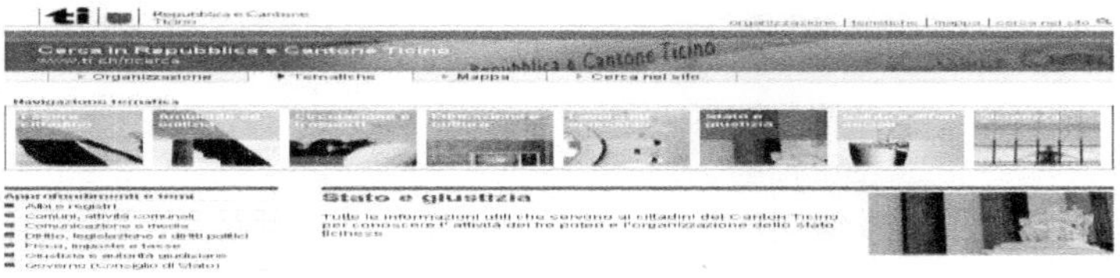

Fig. 4 Subject navigation in the Repubblica e Cantone Ticino web site

In our project, the objective was to provide innovative tools in order to make information access quicker and easier. The "Cantone" site already offered different navigation and searching mechanisms, including:

- subject navigation ("navigazione tematica"), based on a hierarchical classification of categories and subcategories (see figure 4)
- organization navigation ("admnistration"), based on the internal organization of the institution in departments, divisions, etc.
- interactive site map ("mappa")
- traditional search engine ("cerca nel sito") (see figure 5).

In spite of these mechanisms, users still have difficulties in retrieving information of interest. The site has recently undergone a detailed analysis aimed to re-organize the whole information architecture. Within this context, the use of Web 2.0 social tools, and in particular of the tagsonomy approach described above, represents one of the proposed solutions to improve information "scent", i.e. to help people find information they require amongst the huge amount of content available on the site. It is worth noting that the limited use of the web 2.0 social tools is a common trend of most of the corporate web site in Ticino (CH).

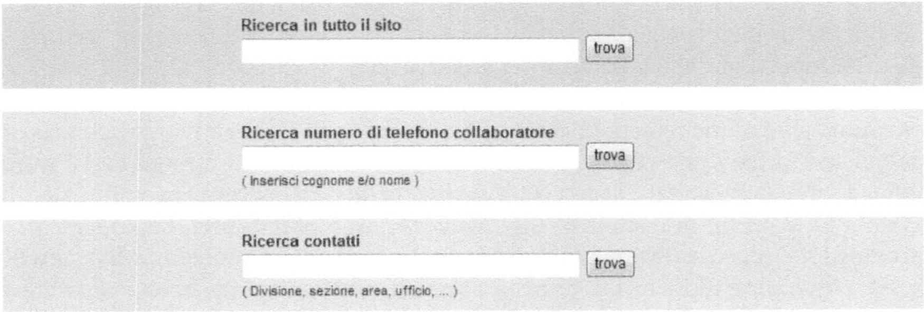

Fig. 5 The web site search engine page without EA

As already highlighted, the EA tagsonomy is visualized as a tag cloud, and used as a tool to support user during the search process (see Fig. 6). This tool is useful as a complement to the search mechanism to directly access information of interest.

Fig. 6 The web site search engine page integrating an EA Tagsonomy Tag Cloud

4.2 Preliminary Evaluation

The tagsonomy approach was initially tested by simulating a number of different scenarios of use, to verify that the system has the expected behavior and to assign proper weights to the parameters used in the formula. Each scenario was simulated considering appropriate user interaction and the results were compared with the previous situation. Examples of the tested scenarios are:

- several users select different tags in the tag cloud during a search session
- the administrator changes the priority of a tag
- the administrator modifies the visibility of a tag.

In these scenarios we have observed for instance the increased size of the tags corresponding to the most visited pages, or the appearance of tags whose priority has been manually incremented by the administrator via the console. In addition, when the visibility of a tag was forbidden by the administrator, it disappeared from the tag cloud leaving space for a new entry. This preliminary test was useful to verify that the EA approach worked properly, and the tagsonomy evolved adaptively in the expected way, changing the tag cloud appearance according to the user input under the administrator governance.

The next step consisted in observing how the tagsonomy evolved in a real situation. One of the remarked issue is that users' most frequently selected tags are associated to the same page, as shown in figure 7. In this case 5 out of 12 words point to the "job search" page (www.ti.ch/concorsi). This result would make the system little useful in a site like the "Cantone" web portal, because no support is provided to retrieve information difficult to find. This problem was partially solved by hiding most of the terms associated to the same page, and showing in the tagcloud only the most frequently searched one. Figure 8 shows the result of this change: only the word "Concorsi" is displayed, removing Lavoro, etc.

This solution was temporarily used to carry on the experiment, but more effective alternatives are necessary. This issue introduces the need for a synonym dictionary, that will be considered for inclusion in a future version.

Further partial data about the number of inserted tags and the number of searches were collected in a first online testing period from March till August 2010. After some improvements to the system, other data were gathered during

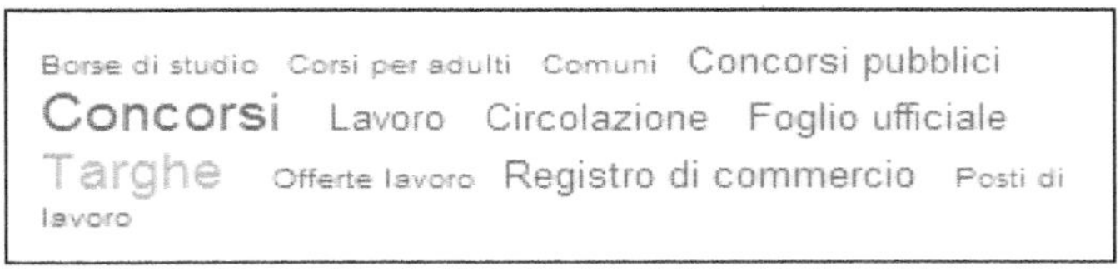

Fig. 7 The initial tagcloud

Fig. 8 The resulting tagcloud after cutting most terms pointing to the same page

September and October 2010. The average number of searches on the web site, which is more than 30.000 per month, has slightly decreased (about 8%) in the last two months. The number of inserted tags has also reasonably decreased from more than 8 thousands in the first month of use (March), to about 4 thousands in July and down to about 2 thousands in October 2010. This decrease confirms the fact that after an initial phase of new entry tags, the tag collection starts to contain some of the most frequently used tags, which do not need to be inserted again, and is more and more converging towards a well established folksonomy.

In addition, analyzing the top 10 searched word from July to October 2010, it was observed that the users' searched words are always the same and the result presented in the tagcloud does not change a lot over time, unless the administrator changes parameters such as the tag priority and visibility. In particular, these words indicate the main subjects of interest for the end user of this institutional web portal, i.e. the citizen. Examples of these words are *concorsi, lavoro, registro di commercio, targhe, circolazione, borse di studio*, etc. Therefore the Easy Access tagcloud provides useful clues in understanding which information users are mostly interested in. This could be also taken into account to change the home page of the site by giving direct access to the most frequently searched words.

The reported evaluation data are only preliminary, considering the EA approach on a relative short period. A longer testing period, over one year or more, is necessary to consolidate the results and to be able to draw more statically based conclusions.

5 Conclusions

The preliminary evaluation results seem to be encouraging from a technical point of view. The tagsonomy module has been made publicly available on the considered web site since summer 2010. Although some conclusions can already be drawn from this initial usage period, additional tests are on going.

At this stage of development it is possible to draw some interesting findings. The EA generated tag cloud is a flexible tool to help users find information of interest because it can reflect the user terminology point of view, and, at the same time, takes into account the top-down predefined classification.

In addition, tags, automatically provided by users through the folksonomy, could be used to provide suggestions about a possible reorganization of the site information architecture (organization of menus, labels, etc.). The tag cloud provides direct connections to information other users have associated to specific tags, reducing the number of interactions to reach a specific content, adaptively to time and context.

Another interesting aspect is that users are not required to explicitly insert tags, that are automatically derived from the search keywords, through an efficient controllable algorithm. Finally, a remarkable feature of the EA system is the administration console, a useful tool to manage the tag cloud and configure several parameters.

In conclusion, it is worth noting that the EA tagsonomy approach can be applied to other web sites with similar characteristics such as large and complex amount of information, heterogeneous target user, and high access rate.

Acknowledgments. The Easy Access system has been accomplished as partial fulfillment of Petra Rota's Bachelor degree Thesis at SUPSI, during the period May-Aug. 2009, in collaboration with the "Repubblica e Cantone Ticino". Many thanks to "Cancelleria dello Stato, Ufficio della comunicazione elettronica", and, in particular, to Cristina Allegri and all the involved staff.

References

1. Pirolli, P.L.T.: Information Foraging Theory: Adaptive Interaction with Information. Oxford Series in Human-Technology Interaction. Oxford University Press, Oxford (2007)
2. Nielsen, J., Loranger, H.: Prioritizing Web Usability. New Riders Press, Berkeley (2006)
3. SearchTools.com: Tools for Taxonomies, Browsable Directories, and Classifying Documents into Categories, `http://www.searchtools.com/info/classifiers-tools.html` (accessed December 2009)
4. Drupal, Organizing content with taxonomy (2010), `http://drupal.org/handbook/modules/taxonomy` (accessed November 2010)
5. Carcillo, F., Rosati, L.: Tags for citizens: Integrating top-down and bottom-up classification in the turin municipality website. In: Schuler, D. (ed.) HCII 2007 and OCSC 2007. LNCS, vol. 4564, pp. 256–264. Springer, Heidelberg (2007)
6. Trant, J.: Studying Social Tagging and Folksonomy: A Review and Framework. Journal of Digital Information 10(1) (2009)

7. Quintarelli, E., Resmini, A., Rosati, L.: Facetag. In: Integrating Bottom-up and Top-down Classification in a Social Tagging System, Information Architecture Summit, Las Vegas (2007)
8. Loasby, K.: Changing Approaches to Metadata at bbc.co.uk: From Chaos to Control and Then Letting Go Again, in ASIS&T Bulletin (October/November 2006), `http://www.asis.org/Bulletin/Oct-06/loasby.html`
9. Rota, P.: Easy Access - Ricerca ed accesso intuitivo ad informazioni per portali Web, Bachelor thesis, SUPSI DTI, Manno (Lugano-CH) (September 2009)

Conceptual Query Expansion and Visual Search Results Exploration for Web Image Retrieval

Enamul Hoque, Grant Strong, Orland Hoeber, and Minglun Gong

Abstract. Most approaches to image retrieval on the Web have their basis in document search techniques. Images are indexed based on the text that is related to the images. Queries are matched to this text to produce a set of search results, which are organized in paged grids that are reminiscent of lists of documents. Due to ambiguity both with the user-supplied query and with the text used to describe the images within the search index, most image searches contain many irrelevant images distributed throughout the search results. In this paper we present a method for addressing this problem. We perform conceptual query expansion using Wikipedia in order to generate a diverse range of images for each query, and then use a multi-resolution self organizing map to group visually similar images. The resulting interface acts as an intelligent search assistant, automatically diversifying the search results and then allowing the searcher to interactively highlight and filter images based on the concepts, and zoom into an area within the image space to show additional images that are visually similar.

Keywords: conceptual query expansion, image search results organization, web image retrieval, interactive exploration.

1 Introduction

Web image retrieval has traditionally followed an approach that is an extension of Web document search techniques [12]. The textual information surrounding and associated with a particular image is used as the core information that describes the image. Using such textual information allows Web search engines to leverage their

Enamul Hoque · Grant Strong · Orland Hoeber · Minglun Gong
Department of Computer Science,
Memorial University,
St. John's, NL, A1B 3X5, Canada
e-mail: enamulp@mun.ca, strong@cs.mun.ca
hoeber@cs.mun.ca, gong@cs.mun.ca

E. Mugellini et al. (Eds.): Advances in Intelligent Web Mastering - 3, AISC 86, pp. 73–82.
springerlink.com

existing expertise and infrastructure associated with searching for documents. This approach can work well when images are concisely and accurately described within Web pages, and when searchers provide accurate descriptions of their image needs. Unfortunately, the accuracy of the descriptions given to images on the Web cannot be enforced; nor can we rely upon searchers providing clear textual descriptions of the images they seek. When these conditions are not met, the search results may include many non-relevant images.

Recent studies on user behaviour with respect to image search have found that queries are often very short [1]. The difficulty with short queries is that they can be open to many different interpretations. It is possible that different searchers may enter the same query, but their intentions and needs may vary significantly from each other. In some cases, the searcher may provide a short query because they wish to see a broad range of images associated with their query; in other cases, the short query may include all of the information the searcher can recall at the moment. Simply matching these queries to the information available for the collection of images, and providing a paged grid of image search results may not be an effective approach for image retrieval.

It has been noted that many image search queries are associated with conceptual domains that include proper nouns (i.e., people's names and locations) [1, 9]. Searches of this type will often have a specific focus, but a less specific aim with respect to resolving the image need (i.e., many different images may be viewed as relevant). In situations such as this, it may be beneficial to promote diversity in the image search results, and then allow the searcher to explore within the collection. The goal of our research is to support such an approach to image search.

In this paper we present a conceptual query expansion method combined with a neural network based image organization technique in order to provide a highly diversified collection of images and an interactive interface for assisting searchers to conceptually and visually explore the image search space. Our system automatically extracts a list of concepts from Wikipedia that are related to the query. These concepts are used as the source of the query expansion to retrieve a broad range of images. The images are visually clustered using a similarity-based approach that employs a multi-resolution extension to self organizing maps (SOM). Concepts used for query expansion are also presented to the searcher in a hierarchical manner based on a conceptual ontology.

The benefit of this approach is that it allows searchers to begin with short and ambiguous queries, which are automatically expanded to provide a diverse range of images. The searcher can then browse the hierarchy of concepts that produced the expanded queries, focusing on those that provide a more accurate description of their needs and filtering based on those that are not relevant. In addition, the searcher can visually explore the search results space, zooming into an area that includes images that look like what they are seeking. Used together, these operations empower the searcher to take an active role in the image retrieval process, supporting their ability to refine their image needs as they explore the image search space.

2 Related Work

Many Web image search approaches are based on traditional keyword based query formulation and matching [12]. Within these, only the text that is associated with the image is considered, without taking advantage of the featuers of the image itself. To address the limitations of this approach, content based image retrieval (CBIR) has been studied as a method for determining image similarity based on low level visual features [4]. However, such approaches can lead to a *semantic gap*: the gap between the way a person finds similarities between images at the conceptual level and the way the system generates similarity based on pixel statistics [5].

One of the recent trends in Web image retrieval employs traditional document search technique combined with CBIR to organize the results with the goal of diversification. In work that was a pre-cursor to Google Swirl, a similarity graph is generated among the results of a textual query using visual features [10]. The PageRank algorithm is applied to the similarity graph to select the authority nodes as the representative images. However it does not analyze semantically related concepts of the textual query; rather it only considers visual features for diversifying the results.

A promising direction for improving the quality of search result in general is the introduction of query expansion based on concepts related to the query [6]. Such an approach is particularly useful for diversifying the search results, and can enable searchers to assist with the query refinement process. However there are a number of challenges associated with conceptual query expansion. The first problem is finding a suitable knowledge base that has sufficient coverage of a realistic conceptual domain. While others have used WordNet to enhance image retrieval [11], it does not contain information pertaining to the proper nouns that are common in image search queries. Using Wikipedia for reformulating queries has shown promise [16], and is the approach we use in our work.

The second challenge is in ranking the extracted concepts for the purposes of selecting the most relevant of these. A useful approach to this problem is to measure the semantic relatedness between the original query and each of the concepts derived from that query. A number of different methods have been devised to use Wikipedia for this purpose, including WikiRelate! [20], Explicit Semantic Analysis (ESA) [7], and Wikipedia Link-based Measure (WLM) [13]. Due to the computationally efficiency and accuracy of WLM, we use this approach in our work.

The last challenge when using conceptual query expansion for the purposes of image search is in the organization of the resulting images. By expanding the query, the number of images within the search results can grow very large. Further, the rank of the search results from a particular query may not be as important as the visual features of the images. As such, a useful approach is to take advantage of the visual features of the images when organizing the search results, and then allow the searcher to browse and explore within the search results space.

Similarity-based image browsing (SBIB) is an approach that takes advantage of the fundamental aspects of CBIR, but eliminates the need for the searcher to identify a set of relevant images *a priori*. Images are organized based solely on their features, allowing searchers to explore the collection even if they do not have a clearly defined information need [8]. The challenge of SBIB is to arrange images based on visual similarities in such a way as to support the browsing and exploration experience. While a number of different approaches have been proposed in the literature [8], we use a method that employs a novel multi-resolution SOM. This approach provides both an organizing structure for the images and a measure of importance that can be used to determine which images to display when space is limited [17, 18]. The interactive features within this approach have been shown to be very useful and easy to use [19].

3 Conceptual Query Expansion for Image Search

One of the main features of this work is the method by which the image search query is automatically expanded. For the short and ambiguous queries that are common in image search, query expansion attempts to capture the various aspects of the query. The objective is to diversify the range of images retrieved, providing a broad view of what is available. The problems of then allowing the searcher to narrow down the image search results and focus on the aspects that match their particular needs are addressed in Section 4.

The process of performing conceptual query expansion of image search queries follows three steps: extracting concepts from Wikipedia, ranking the extracted concepts, and generating the expanded queries. While others have used Wikipedia for query expansion in the context of general Web search [15], the approach we use is novel in that it takes advantage of specific aspects of image search. The details are explained in the remainder of this section.

3.1 Extracting Concepts from Wikipedia

For this work, we use Wikipedia as the core knowledge base for the query expansion process. Wikipedia is an excellent source of information for the purposes of image search since it includes a large number of articles describing people, places, landmarks, animals, and plants. It is densely structured, with hundreds of millions of links between articles within the knowledge base. Most of the articles also contain various representative images and associated textual captions.

A dump of the Wikipedia collection was obtained in June 2010, and was preprocessed to support the type of knowledge extraction required for our purposes. In analyzing Wikipedia, we observed that the in-going links (i.e., articles that are linked to the current article) and out-going links (i.e., articles to which the current article links) of an article often provide meaningful information that is closely related to the concept upon which the article is written. Therefore, for each article (concept) within the collection, the in-going and out-going links were detected and extracted as related concepts.

We also found that the captions surrounding the images present within a given article can often provide a valuable perspective on the visual features associated with the article. To ensure the inclusion of all relevant concepts associated with the image captions, we use Wikifier [14] to augment the captions with links to relevant Wikipedia articles that may have been missed by the author of the article, and use these links to extract their associated concepts.

Matching a user-supplied query to this knowledge base is simply a matter of selecting the best matching article (referred to as the home article) from Wikipedia using its search feature. In the case where the query is highly ambiguous and Wikipedia returns multiple articles, the user is prompted to select the home article that is the closest match for their information need. While it is possible for a searcher to enter a query that does not include any matches within Wikipedia, this is highly unlikely given the type of information people commonly look for with image search systems, and how closely this matches the type of information present in Wikipedia (i.e., people, places, and things).

The end result of this process is the selection of the home article, along with a list of all the other articles that are part of the in-going or out-going links for the home article, and the articles that originate from the captions of the images within the home article. These concepts provide the basis for automatic query expansion process.

3.2 *Ranking the Extracted Concepts*

Due to the richness of Wikipedia, the number of concepts obtained in the process described above may become very large. While it is good that so much information is available for the query expansion process, there is a risk in expanding the query too broadly. In order to address this potential problem, we rank the extracted concepts and use only those that are most similar to the home article.

For each of the candidate articles $\{c_i | 1 \leq i \leq C\}$ extracted from the home article, a semantic relatedness measure is applied between the home article h and the extracted articles. WLM [13] is used for this purpose, taking advantage of the hyperlink structure of the associated articles to find out how much they share in common. In order to give preference to the concepts that have been extracted from the image captions within the home article, we use a re-weighting function to determine the relatedness score:

$$r(c_i, h) = \min(WLM(c_i, h)(1 + \alpha), 1)$$

Since WLM provides a value in the [0,1] range, we ensure that the relatedness score remains in this range with the min function. The re-weighting factor α is provided according to the following function:

$$\alpha = \begin{cases} k\frac{C}{N} & \text{if concept } c_i \text{ originates from a caption} \\ 0 & \text{otherwise} \end{cases}$$

Here, C is the number of concepts that have been extracted from the home article, and N is the number of concepts to be selected for query expansion, and k is a system parameter that controls the importance of the concepts derived from the captions. In our prototype implementation $N = 30$, $k = 0.01$, and C commonly ranges from 300 to 600. This results in a 10 - 20% increase in the score for the concepts derived from the captions, with proportionally more importance being given when there are more concepts extracted from the home article.

The outcome of this process is that the top-N concepts are selected from among the candidate articles, such that those which came from the image captions are given preference over those which came from in-going and out-going links of the home article. These concepts are used as the source for the query expansion.

3.3 Generating Expanded Queries

In order to ensure that the expanded queries remain focused on the topic of the query itself, the original query Q is pre-pended to each of the top-N related concepts $\{c_r | 0 \leq r \leq N\}$ as $< Q, c_r >$. We define c_0 to be null, producing the original query plus N expanded queries.

Given that individual expanded queries have differing degrees of relevance to the original query, we dynamically determine how many images to retrieve for each expanded query based on their relatedness score to the home article:

$$I_r = \frac{r(c_r, h) \times I_t}{\sum_{k=0}^{N} r(c_k, h)}$$

Here, r is the same function used to generate the relatedness score in the concept ranking process, and I_t is the total number of images to be retrieved by all of the queries (we set $I_t = 300$ in the current prototype). Since the null expanded query (c_0) is the original query, we define $r(c_0, h) = 1$ in the above calculation. All of the queries are sent to the Google AJAX Search, and the desired number of images are retrieved. Duplicate images are deleted based on the URL of the source image (as provided by the underlying search engine).

4 Visual and Conceptual Search Results Exploration

The difficulty with retrieving a broad and diversified range of images is how to then present these in a way that allows the searcher to focus on the specific aspect of the query they are interested in. A naïve approach would be to use a traditional paged grid layout of the images, ordered by their rank in the search results list and perhaps the semantic relatedness between their source concept to the original query concept. However, such an approach may not be all that effective in supporting image search tasks since the meaning of the organization of the images may be rather obscure. Instead, we propose a visual and interactive method for exploring the broad range of images retrieved from the expanded queries, taking advantage of both the visual features of the images and the concept from which they came.

4.1 Multi-resolution SOM-Based Image Organization

In order to organize images based on visual similarity, we must first extract the visual features from the images. While a number of approaches have been studied within the domain of image processing [4], we use color-gradient correlation since it is efficient to calculate and offers good organizational performance [18]. We then train a SOM in a process similar to [17] to organize the images. The topology preserving property of the SOM ensures that images with similar feature vectors are mapped to locations that are closer to each other, and vice versa.

While the trained SOM provides the positioning coordinates for each image, it is impractical to provide and occlusion-free display of all images when the search results collection is large and the display resolution is limited. To facilitate the selection of which images to display under these space constraints, we assign priorities to all images based which are more representative of the images nearby. Of the images in a region of the SOM, a particular image is considered to be more representative if its feature vector is most similar to the average of the feature vectors in that region. This calculation is performed at progressively smaller resolutions, producing in a multi-resolution extension to the SOM [17]. As a result, only images with high display priority are shown when there is insufficient space to display all images. The amount of space available is relative to the screen resolution, as well as two parameters that can be controlled interactively: zoom level and image size.

User evaluations with this framework for visual image browsing and exploration studied the benefits of organizing the images based solely on their similarity (following a messy-desk metaphor) and in a more structured layout (following a neat-desk metaphor) [19]. While this study found that the approach can be very useful in comparison to traditional methods for organizing image search results, the differences between the layouts appeared to be based on personal preferences. For this work, we use the neat-desk layout since it provides less of a departure from what searchers expect in the presentation of image search results, while still maintaining a visual encoding of the degree of similarity.

In order to align the images to a structured grid, we generate a kd-tree [2] using the positions provided by the SOM. At the default zoom level, only the images with the highest display priority are shown. As the user zooms into a particular location, the amount of space available for displaying images grows. To avoid occlusion problems, new images are displayed only when there is sufficient space available between the currently displayed images.

4.2 Concept Hierarchy Focusing and Filtering

In addition to arranging the images from the search results visually, our approach also uses the concepts from which we derived the expanded queries to support focusing and filtering operations. Each of these concepts is mapped to an ontology using DBPedia [3]. This ontology is displayed to the searcher as a hierarchy, with the most semantically similar concepts to the original query placed at the top.

(a) The search results include images from the expanded queries, organized based on their visual similarity. Due to space constraints, only images with a high display priority are shown.

(b) After filtering the search results to remove irrelevant concepts, the search results are focused on the concept of "Bastille Day" which brings those images to the foreground, and zoomed to show the images in greater detail.

Fig. 1 Search results from the query "Eiffel Tower". The images from the expanded query are provided in (a); the search results are filtered, focused, and zoomed in (b).

The searcher can use this hierarchy of concepts for both focusing and filtering operations. By clicking on any of the concepts, all the images that were retrieved as a result of that concept are pulled to the front of the display (temporarily increasing their display priority within the image organization process described above); the remaining images are dimmed giving the focused images more visual prominence.

The searcher can use checkboxes associated with each node in the concept hierarchy to filter the search results, removing the associated images from the display. At any time during the use of these conceptual filtering and focusing operations, the searcher can perform additional visual exploration, zooming into an area of interest to show additional visually similar images. Screenshots of the search results exploration interface are provided in Figure 1.

5 Conclusions and Future Work

In this paper, we describe an approach for performing conceptual query expansion, producing a diversified set of image search results which are then organized based on their visual features, and presented within an interactive interface. The primary contributions of this work are the novel use of Wikipedia for image retrieval, and the interactive support provided for conceptual and visual exploration within the image search space.

Future work includes adding features to support complex multi-concept queries, adding additional features that support interactive query refinement loops and query-by-example, and evaluating the approach through user studies. We are also examining the benefit of including conceptual information within the image organization process.

References

1. André, P., Cutrell, E., Tan, D.S., Smith, G.: Designing novel image search interfaces by understanding unique characteristics and usage. In: Proceedings of the IFIP Conference on Human-Computer Interaction, pp. 340–353 (2009)
2. Bentley, J.L.: Multidimensional binary search trees used for associative searching. Communications of the ACM 18(9), 509–517 (1975)
3. Bizer, C., Lehmann, J., Kobilarov, G., Auer, S., Becker, C., Cyganiak, R., Hellmann, S.: DBpedia - a crystallization point for the web of data. Journal of Web Semantics: Science, Services and Agents on the World Wide Web 7(3), 154–165 (2009)
4. Datta, R., Joshi, D., Li, J., Wang, J.Z.: Image retrieval: Ideas, influences, and trends of the new age. ACM Computing Surveys 40(2), 1–60 (2008)
5. Enser, P., Sandom, C.: Towards a comprehensive survey of the semantic gap in visual image retrieval. In: Bakker, E.M., Lew, M., Huang, T.S., Sebe, N., Zhou, X.S. (eds.) CIVR 2003. LNCS, vol. 2728, pp. 291–299. Springer, Heidelberg (2003)
6. Fonseca, B.M., Golgher, P., Pôssas, B., Ribeiro-Neto, B., Ziviani, N.: Concept-based interactive query expansion. In: Proceedings of the ACM International Conference on Information and Knowledge Management, pp. 696–703 (2005)

7. Gabrilovich, E., Markovitch, S.: Computing semantic relatedness using Wikipedia-based explicit semantic analysis. In: Proceedings of the International Joint Conference on Artificial Intelligence, pp. 1606–1611 (2007)
8. Heesch, D.: A survey of browsing models for content based image retrieval. Multimedia Tools and Applications 42(2), 261–284 (2008)
9. Jansen, B.J., Spink, A., Pedersen, J.: An analysis of multimedia searching on AltaVista. In: Proceedings of the ACM SIGMM International Workshop on Multimedia Information Retrieval, pp. 186–192 (2003)
10. Jing, Y., Baluja, S.: VisualRank: Applying PageRank to large-scale image search. IEEE Transactions on Pattern Analysis and Machine Intelligence 30(11), 1877–1890 (2008)
11. Joshi, D., Datta, R., Zhuang, Z., Weiss, W.P., Friedenberg, M., Li, J., Wang, J.Z.: PARAgrab: A comprehensive architecture for web image management and multimodal querying. In: Proceedings of the International Conference on Very Large Databases, pp. 1163–1166 (2006)
12. Kherfi, M.L., Ziou, D., Bernardi, A.: Image retrieval from the world wide web: Issues, techniques, and systems. ACM Computing Surveys 36(1), 35–67 (2004)
13. Milne, D., Witten, I.H.: An effective, low-cost measure of semantic relatedness obtained from Wikipedia links. In: Proceedings of the AAAI Workshop on Wikipedia and Artificial Intelligence, pp. 25–30 (2008)
14. Milne, D., Witten, I.H.: Learning to link with Wikipedia. In: Proceedings of the ACM Conference on Information and Knowledge Management, pp. 509–518 (2008)
15. Milne, D., Witten, I.H., Nichols, D.M.: A knowledge-based search engine powered by Wikipedia. In: Proceedings of the ACM Conference on Information and Knowledge Management, pp. 445–454 (2007)
16. Myoupo, D., Popescu, A., Le Borgne, H., Moëllic, P.-A.: Multimodal image retrieval over a large database. In: Peters, C., Caputo, B., Gonzalo, J., Jones, G.J.F., Kalpathy-Cramer, J., Müller, H., Tsikrika, T. (eds.) CLEF 2009. LNCS, vol. 6242, pp. 177–184. Springer, Heidelberg (2010)
17. Strong, G., Gong, M.: Browsing a large collection of community photos based on similarity on GPU. In: Proceedings of the International Symposium on Advances in Visual Computing, pp. 390–399 (2008)
18. Strong, G., Gong, M.: Organizing and browsing photos using different feature vectors and their evaluations. In: Proceedings of the International Conference on Image and Video Retrieval, pp. 1–8 (2009)
19. Strong, G., Hoeber, O., Gong, M.: Visual image browsing and exploration (Vibe): User evaluations of image search tasks. In: An, A., Lingras, P., Petty, S., Huang, R. (eds.) AMT 2010. LNCS, vol. 6335, pp. 424–435. Springer, Heidelberg (2010)
20. Strube, M., Ponzetto, S.P.: WikiRelate! computing semantic relatedness using Wikipedia. In: Proceedings of the AAAI Conference on Artificial Intelligence, pp. 1419–1424 (2006)

Memoria-Mea: Combining Semantic Technologies and Interactive Visualization Techniques for Personal Information Management

Francesco Carrino, Maria Sokhn, Elena Mugellini, and Omar Abou Khaled

Abstract. Thanks to the development of new technologies (such as PCs, PDAs, digital cameras, etc.) and with the advent of the Web, people are faced to deal with an increasing amount of information during their every day-life activities. As a consequence, the problem of finding the right information, at the right time, in a short period of time, becomes more and more crucial. Several technologies such as semantic web data mining and interactive visualizations are nowadays available to address these specific issues. However they have not been jointly exploited. With this respect, the paper presents a novel approach. Finally a prototype that validate our approach is presented.

Keywords: Personal Information Manager, Ontology, Data Mining, Information Retrieval, Interactive Visualization.

1 Introduction

With the constant progress in information retrieval and storage, the amount of information a person owns and handles is drastically increasing. As a consequence every day information retrieval activities become more and more time-consuming and less efficient. Several technologies already exist to address these issues, such as semantic annotations to facilitate information retrieval, ontology and inference engine to reason about data, interactive visualization techniques to help the user browsing and navigating within his personal collection of information.

Francesco Carrino · Maria Sokhn · Elena Mugellini · Omar Abou Khaled
University of Applied Sciences of Fribourg,
Boulevard Perolles, 80, 1700, Fribourg
e-mail: `francesco.carrino@hefr.ch`, `maria.sokhn@hefr.ch`
`elena.mugellini@hefr.ch`, `omar.aboukhaled@hefr.ch`

E. Mugellini et al. (Eds.): Advances in Intelligent Web Mastering - 3, AISC 86, pp. 83–92.
springerlink.com

Semantic web technologies provide new powerful ways of handling data. Moreover semantic to data gives to machines new possibilities of reasoning about data and deducting new, unstated, information. In this context, usual knowledge management systems database can be enriched by new ways of describing, linking, storing and treating data.

The use of multiple visualization techniques, adapted to different visualization needs, can improve information browsing and retrieval activities towards a more attractive user experience.

What we often miss is an integrated system which jointly takes advantage of these technologies to provide the user with the right information, at the right time. This paper presents Memoria-Mea [1], a novel approach to address this issue; in order to validate the concept, the paper presents as prototype an ontology-based annotation tool along with a reasoning engine and a graphical user interface providing interactive visualizations of multimedia data in order to facilitate user search and browsing activities.

The paper is organized as follow: section 2 provides an overview of relevant state-of-the-art, section 3 describes the Memoria-Mea project from a logical point of view, section 4 describes the technical architecture, section 5 presents the prototype that validate the system, and section 6 concludes the paper.

2 Related Work

For the sake of simplicity in this paragraph we will focus only on existing works which aim at combining those technologies into an integrated system to facilitate information management of personal collection of data.

Numerous projects are undergoing in the domain of semantic information and knowledge management application. One of the most important is Nepomuk project [6], [7]. Nepomuk is a FP6 European project which intends to develop a semantic social desktop for extending the personal desktop into a networked collaboration environment. This project integrates some open source tools such as Gnowsis [8] a semantic desktop framework, and Aperture [9] a java framework for extracting and querying full-text content and metadata from various information systems. While Nepomuk focuses on social aspects, our work deals mainly with issues related to visualization and management of personal collection of information.

Another interesting project in multimedia content management domain is the FP6 AceMedia project [10]. AceMedia is user-centered tool focusing on the concept of knowledge assisted and adaptive multimedia content management. It explores and integrates various technologies such as automatic annotation of multimedia content, content classification, person/face detection and recognition, as well as multilingual and ontological text analysis. Our work goes in the same idea of integration, furthermore providing a virtual queries system dedicated to interrogate different sort of annotated data.

Finally a further interesting project is Stuff I've Seen (SIS) [11]. The main goal of the project is to make it easy to find information you've seen before (whether it came as email, attachments, files, web pages, appointments, etc.). Moreover it proposes some innovative visualization techniques based on the concept of Memory Landmarks which have been integrated in our work. Contrary to our work, no integration of ontology or semantic web technologies is provided.

3 Memoria-Mea: Logical Architecture

Memoria-Mea project aims to develop a Personal Information Management system (PIM) for managing multimedia content.

The goal of the project is to develop a system supporting the user in searching, browsing and visualizing "multimedia memories" (e.g. pictures, videos, audio file, text file, mobile phone media, etc.). Memoria-Mea is based on personalized information indexing and classification techniques based on a rich semantic model which integrates both ontology-based annotations and reasoning along with data mining techniques.

In particular we identified seven levels in this logical architecture. The first one called raw data[1] represents all multimedia files (text files, sound, video, image, web pages, e-mail and so on) which constitute what we call a personal collection of information. These raw data are supposed to be stored on the user's personal computer.

The second level - automatic / manual annotation - represents the data annotation and indexing activities using automatic, semi-automatic or manual annotation tools. At this level we adopt a twofold strategy. On the one hand, we integrate, as much as possible, existing tools such as Google Desktop Search (GDS), for automatic indexing, and Protégé, for manual ontology edition and annotation. On the other hand, we develop our own annotation tools such as *Facia Mea*, a graphical media annotation application for mobile devices, or *PersoMemo*, an intuitive application for annotation, search and visualization of personal information, or *MemoSAM* (contraction of "Memoria-Mea Semantic Annotation of Multimedia", see section 5.2, an ontology driven annotation tool to assist user's annotation of personal resources according to predefined domain ontology.

The described data level stores data annotated at the previous level. These data are then enriched via data mining as well as ontology processing (intelligent data processing level) to reach the fifth level, intelligent indexed data.

The last two levels called intelligent search engine and data exploitation offer to the user several interfaces to search, browse and visualize information. The intelligent search engine hides to the user all the complexity of the underlying processing mechanisms.

4 Memoria-Mea: Technical Architecture

From a technical point of view (see Figure 2), the Memoria-Mea architecture is composed by four modules communicate with each other, on the top of a layer containing different forms of data (raw data, semantic database, classical database, etc.):

[1] For *raw data* we mean data that have been collected by the user but which have not been semantic annotated nor data mining processed yet.

- The **Visualization Module** allows the user to search, browse and visualize information. This module is implemented by the application *MemoSIV* (contraction of "Memoria-Mea Smart Information Visualization") that represents also the highest layer in the logical view presented above (Figure 1).
- The **Annotation Module** manages three different data levels: without annotation, with simple annotation, with semantic annotation. The first one can be see how the mere integration of tools for the automatic annotation such as GDS. The second level is represented by integrated tools that allow to add information, categorize data in various ways, etc. In this level we have developed several applications as *PersoMemo* and *FaciaMea* mentioned above. The third and most refined way to do annotation integrated in the Memoria-Mea platform is represented bye the *MemoSAM* prototype (see the section 5.2); it can store semantic data, reason about them, infer new knowledge. This application adds semantic information to data and link them together using web semantic technologies. As a knowledge base, it contains the vocabularies defined by different ontologies, the basic data (assertions) provided by different modules, and also all the inferred data that can be deduced by a reasoning engine.
- The **Virtual Queries Module** aims to allow the user to formulate queries in a natural way, without concerning about the database type behind. Each query is analyzed and processed in order to extract the maximum information from all kind of database (semantic or not). For instance, in the travel context, an user

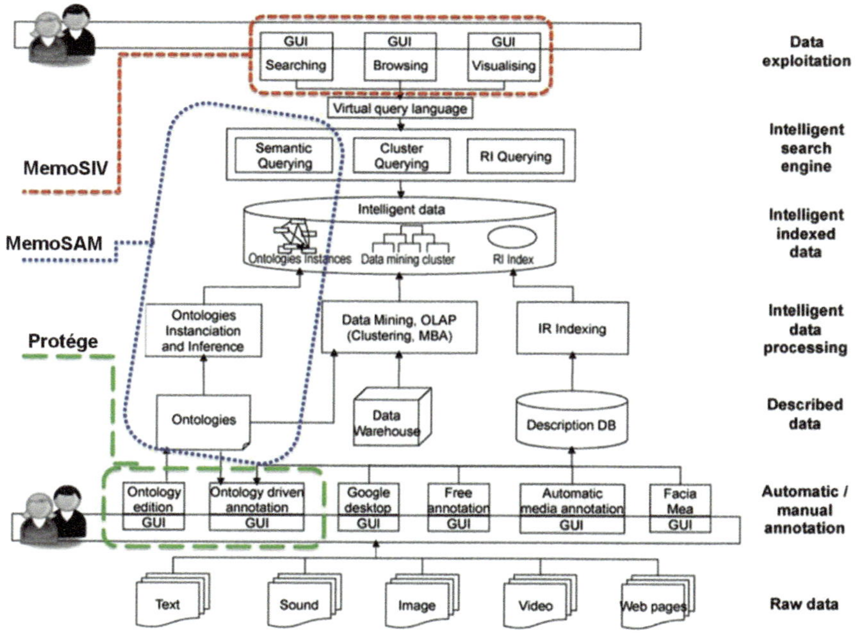

Fig. 1 Logical architecture of Memoria Mea

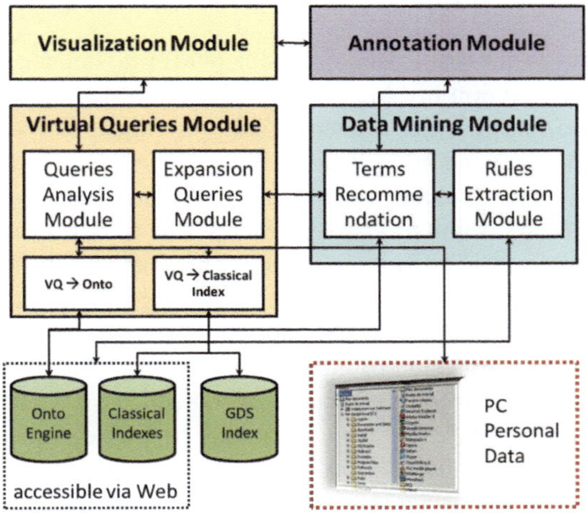

Fig. 2 Technical Architecture of Memoria-Maa

that asks about his *memories* concerning “Paris”, he will get data about the city and the other concept semantically related to it (e.g. administrative regions, countries, etc.) when this information is made available by the semantic annotation; otherwise more simple research as Google Desktop Search will be preferred.

- The **Data Mining Module** is composed by two sub-module: Terms Recommendation Module - with the task of analyzing the data present in the database giving how output the rules of association or of co-occurrence between terms; Rules Extraction Module - this module is called from both Annotation Module and Virtual Queries Module. This, in turn, is divided in three submodules that use different techniques to make a recommendation: the submodule based on annotations; the submodule based on queries; the submodule based on ontologies.

A prototype validating the Visualization and the Annotation modules will be presented in the next section.

5 Prototype

The prototype[2] developed implements the whole process from ontology creation and ontology-based annotation steps up to the semantic knowledge based creation and querying and interactive visualization process (red, blue and green circled regions on Figure 1). It is presented in the following sections.

[2] Our testbed data are focused on traveling as a privileged application domain since this activity is strongly related to the concept of personal information collection.

5.1 Visualization Module: MemoSIV

The visualization system in the Memoria-Mea platform is managed by the application *MemoSIV* (contraction of "Memoria-Mea Smart Information Visualization"). MemoSIV application provides a user-friendly graphical interface to search and browse within a heterogeneous collection of multimedia content (such as videos, pictures, audio files, textual documents, etc.). When displaying a document for search purpose it is important to visualize its relevant features to facilitate retrieval activity. So far, several visualization techniques have already been developed to visualize specific categories of document (e.g. providing a thumbnail to visualize pictures, showing the title to visualize textual documents, using memory landmarks to facilitate information retrieval [11], etc.). Based on this observation, our main idea behind the design of MemoSIV interface was to seamlessly combine those visualization techniques in a single application in order to provide the user with different visualizations metaphors, adapted to the different media to be displayed, thus facilitating information browsing and retrieval activity. According to this, the main interface of MemoSIV application provides two main navigation metaphors:

- The first one, rather traditional, allows the user to browse his/her data according to their type (e.g. textual document, videos, audio files, etc.).
- The second one allows the user to browse his/her data according to some main events, chronologically ordered (e.g. the main stages of a travel).

The user can at any moment swing from one visualization to the other one, browsing information by association, like our human *memory* does. Figure 3 shows the main interface of the application, when the user selects one particular topic ("theme", at the center of the circles) of interest (for example the travel in Morocco, as show in Figure 3), the system displays related content according to the two main navigation metaphors.

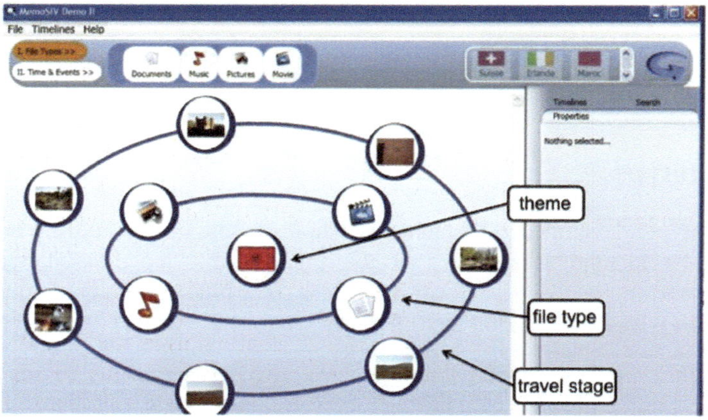

Fig. 3 MemoSIV main interface

The first one, represented by the inner circle in Figure 3, is called “File Types”. It is based on the “type” of multimedia content (each bubble represents one of the following media types: pictures, videos, audio and textual documents). The second one, represented by the outer circle in Figure 3, is called “Time & Events”. It is based on main events (“stage”) related to the selected theme (for instance in the case of a travel in Morocco each bubble could represent the main cities the person visited during the vacation). Alternatively these two navigation metaphors can be selected by clicking on their corresponding buttons on the top-left of the interface.

From the main interface the user can either directly query the system to find a particular resource (search command on the right) or just browse the different media simply by clicking on one of the bubble on the two circles.

Different visualizers are implemented for each data type. Thus Textual documents, Pictures, Videos, Musics have an interface dedicated which allow the user to search, visualize or filter information in a specific way according to the data type. A special viewer is given for the Time & Events visualization.

As mentioned above, MemoSIV has to managed three different levels of “smart” data: data without any annotation (e.g. annotated automatically by Google Desktop Search), data with simple annotation (e.g. from FaciaMea or PersoMemo) and data with semantic annotation (e.g. previously annotated with MemoSAM, see 5.2).

To exploit this extra information a special viewer will be made available in the next version (see 6) where linked concept will be showed following well known methods showed in works as [28] and [29]. Also the *MemoSAM Query Builder* will be integrated (see section 5.2) in the Virtual Queries Module.

5.2 Annotation Module: MemoSAM

MemoSAM is a semantic knowledge base engine. It can store semantic data, reason about them, infer new knowledge, and give full access to these data to the other Memoria-Mea modules. MemoSAM has no revolutionary semantic web concepts; it is the result of a work on development and integration of semantic technologies to construct a complete semantic engine from storing (insert, update and remove data) to querying. Querying can be done directly in SPARQL [25], through web services or via a Query Builder (that translate simple and user friendly searching criteria into SPARQL queries).

MemoSAM adds semantic information to data and link them together using web semantic technologies. As a knowledge base, it contains the vocabularies defined by different ontologies, the basic data (assertions) provided by different modules, and also all the inferred data that can be deduced by the reasoning engine[3].

MemoSAM is developed in Java and it offers several services accessible through an embedded web server. These services allow mainly to manage the knowledge

[3] Particularly, to do inference we adopted Pellet[27] an OWL 2 reasoner that incorporates optimizations for nominals, conjunctive query answering, and incremental reasoning. Since Memoria-Mea makes use of several ontologies integrated in the same structure, Pellet absolves also the coherence controller function, preventing from initializing the application when the structure is not consistent.

base and to make queries. MemoSAM uses Jena's framework to handle ontologies (definition of data), data (instances of the ontologies) and inferences (new deduction based on rules). Inspired by Joseki's architecture [15], it implements Jetty [16] as a light web Server and makes its content accessible through different web services; between the components of Jena, MemoSAM uses TDB that provides for large scale storage and query of RDF datasets using a pure Java engine. Other semantic tools or frameworks exist to build semantic application (e.g. KAON [17], Sesame [18], Kowari [19], Mulgara [20]). However Jena is one of today's most powerful and widespread tools for semantic development. It offers a large set of well-documented API and supports different rule-based inference engines. This open source framework satisfies our needs in term of flexibility, integration, performance and lightness. Moreover Jena is easily plugged-in with other component and extensible.

MemoSAM considers three conceptual categories of semantic information stored in ontologies: description information, domain information and personal information. Description information contains all information on multimedia files descriptions and their annotation. For example for a photo it can be data concerning the name of the file, its creation date, GPS coordinates, annotations (such as people who appears on this photo), express of interest and comments. Domain information expresses more general concept such as locations or tourism or travel ontology. Personal information deals with user model information (e.g. behavior, preferences, etc.).

MemoSAM contains its own ontology called "MemOnto" which includes description ontology specific to Memoria-Mea needs. It also links other public ontologies and knowledge bases. In order to describe people we integrate the FOAF 'Friend Of A Friend' ontology [21]. Even if FOAF is not a standard in the sense of ISO Standardization, or W3C Process, it depends heavily on W3C's standards work. Mainly FOAF is a simple, well-known and widely used ontology.

Furthermore, to provide location's information we integrate GeoNames [23] website. The GeoNames Ontology is an interesting and large initiative to make it possible to add geospatial semantic information to the Word Wide Web. Geonames claims over 6.5 million unique features whereof 2.2 million populated places and 1.8 million alternate names. The data is accessible through a number of web services and a daily database export. GeoNames is already serving up to over 11 million web service requests per day.

MemoSAM is designed to make ontologies transparent to the other modules of Memoria-Mea. Data exchanges between MemoSAM and other different modules are done using an XML-based common language, thus the other modules don't need to support semantic languages. Through web services MemoSAM gives access to its knowledge base to any software, independently of the language in which it is written.

Demonstration videos and screenshots of MemoSAM are available on the Memoria-Mea website [1].

6 Conclusion and Future Works

This paper presented the Memoria-Mea architecture, a novel approach that combines semantic annotation, semantic knowledge search engine and interactive visualizations techniques to enrich user experience for searching and browsing through personal multimedia data. In addition a prototype that validate the concept has been presented.

As next steps we plan to continue the integration process centered with the introduction of the Virtual Queries Module and the Data Mining Module that will add intelligence to the system. In this direction, MemoSIV will be enhanced since the management of Virtual Query will allow a simplification in the search interface. In the visualization side we are improving the system making possible to automatically generated the interface (menu, filters, etc.) from the ontology (feature already available in the MemoSAM interface) and introducing an interface based on the work of [29] to take advantage from the expressiveness of annotated semantic data.

Finally an accurate evaluation of the prototype will be performed in order to assess how acceptable is for end users to use it. The evaluation will be centered on two main aspects: the ergonomics and usability of the proposed systems and the willingness to use such technology [26].

References

1. Memoria-Mea project, http://www.memoria-mea.ch
2. Popov, B., Kiryakov, A., Ognyanoff, D., Manov, D., Kirilov, A.: KIM - a semantic platform for information extraction and retrieval. Natural Language Engineering 10(3-4), 375–392 (2004)
3. Bertini, M., Del Bimbo, A., Torniai, C., Cucchiara, R., Grana, C.: MOM: multimedia ontology manager. A framework for automatic annotation and semantic retrieval of video sequences. ACM Multimedia, 787–788 (2006)
4. Semantic web tools, http://esw.w3.org/topic/SemanticWebTools
5. Ontology list, http://www.schemaweb.info/default.aspx
6. Groza, T., Handschuh, S., Moeller, K., Grimnes, G., Sauermann, L., Minack, E., Mesnage, C., Jazayeri, M., Reif, G., Gudjonsdottir, R.: The NEPOMUK Project - On the way to the Social Semantic Desktop. In: Proceedings of I-Semantics 2007, pp. 201–211 (2007)
7. Nepomuk project, http://nepomuk.semanticdesktop.org
8. Sauermann, L., Aastrand Grimnes, G., Kiesel, M., Fluit, C., Maus, H., Heim, D., Nadeem, D., Horak, B., Dengel, A.: Semantic desktop 2.0: The gnowsis experience. In: Cruz, I., Decker, S., Allemang, D., Preist, C., Schwabe, D., Mika, P., Uschold, M., Aroyo, L.M. (eds.) ISWC 2006. LNCS, vol. 4273, pp. 887–900. Springer, Heidelberg (2006)
9. Aperture: a Java framework for getting data and metadata, http://aperture.sourceforge.net
10. Kompatsiaris, I., Avrithis, Y., Hobson, P., Strintzis, M.G.: Integrating Knowledge, Semantics and Content for User-Centred Intelligent Media Services: the aceMedia Project. In: WIAMIS 2004, Lisboa, Portugal, April 23 (2004)

11. Dumais, S., Cutrell, E., Cadiz, J.J., Jancke, G., Sarin, R., Robbins, D.C.: Stuff I've Seen: A System for Personal Information Retrieval and Re-Use. In: Proceedings of SIGIR 2003, Toronto, Canada (2003)
12. Google Desktop, http://desktop.google.com
13. Protégé, http://protege.stanford.edu
14. PhotoStuff - An Image Annotation Tool for the Semantic Web, http://www.mindswap.org/2003/PhotoStuff
15. Joseki - A SPARQL Server for Jena, http://www.joseki.org
16. Jetty://, http://jetty.mortbay.org/index.html
17. KAON, http://kaon.semanticweb.org
18. Sesame 2.0, http://www.openrdf.org
19. Wood, D., Gearon, P., Adams, T.: Kowari: A Platform for Semantic Web Storage and Analysis. In: WWW 2005, Chiba, Japan, May 10-14 (2005)
20. Mulgara, http://www.mulgara.org
21. The Friend of a Friend (FOAF) project, http://www.foaf-project.org
22. RELATIONSHIP: A vocabulary for describing relationships between people, http://vocab.org/relationship
23. GeoNames, http://www.geonames.org
24. Huynh, D., Drucker, S., Baudisch, P., Wong, C.: Time Quilt: Scaling up Zoomable Photo Browsers for Large, Unstructured Photo Collections. In: Proceedings of CHI 2005, Portland, USA, pp. 1937–1940 (April 2005)
25. SPARQL, http://www.w3.org/TR/rdf-sparql-query
26. Mavrommati, I., Kameas, A., Markopoulos, P.: An editing tool that manages devices associations in an in-home environment. Personal and Ubiquitous Computing 8(3-4), 255–263 (2004)
27. PELLET, http://clarkparsia.com/pellet
28. Sokhn, M., et al.: Conference knowledge modeling for conference-video-recordings querying & visualization. In: Pro. International Conference on Management of Emergent Digital EcoSystems (2009)
29. Keim, D.A.: Information visualization and visual data mining. IEEE Transactions on Visualization and Computer Graphics 8, 18 (2002)

Cylindric Extensions of Fuzzy Sets. An Application to Linguistic Summarization of Data

Adam Niewiadomski

Abstract. The paper presents remarks on possible using fuzzy sets and their cylindric extensions in representations of compound linguistic terms, e.g. *young and tall*. A description of such a representation is proposed to formalize unions and intersections of fuzzy sets in different universes of discourse but concerning the same object that manifests some attributes. We show an application of the proposed description in linguistic summaries of databases in the sense of Yager.

1 Introduction

The scope of the paper is to present an original description of representation of compound linguistic expressions in terms of cylindric extensions of fuzzy sets. The new description can be especially useful when two or more fuzzy sets are defined in different universes of discourse but the terms represented concern the same object, e.g. *a worker who is young and tall and has average salary*. Performing simple operations of union and intersection of fuzzy sets, cf. [12], representing these imprecise expressions is not possible because *young*, *tall* and *average salary* are defined in different spaces. Therefore, we propose a simple but useful technical means based on the cylindric extension of a fuzzy set to the Cartesian product of those different universes of discourse.

Thus, in the two following subsections of this introduction, we present crucial concepts of cylindric extensions of fuzzy sets and linguistic summaries of databases, respectively. In Section 2, we show representations of compound linguistic terms via cylindric extensions of fuzzy sets. Then, we show applications of cylindric extensions to three chosen aspects of linguistic summarization, Sections 3–5. Finally, we draw some conclusions and further work directions in Section 6.

Adam Niewiadomski
Institute of Information Technology, Technical University of Łódź,
ul. Wólczańska 215 90-924 Łódź, Poland
e-mail: aniewiadomski@ics.p.lodz.pl

E. Mugellini et al. (Eds.): Advances in Intelligent Web Mastering - 3, AISC 86, pp. 93–100.
springerlink.com

1.1 The Definitions of Fuzzy Sets and Their Cylindric Extensions

A *fuzzy set* A in a non-empty universe of discourse $\mathscr{X}$, is defined [12] as

$$A = \{\langle x, \mu_A(x)\rangle : x \in \mathscr{X}\} \tag{1}$$

where $\mu_A : \mathscr{X} \to [0,1]$ is *membership function* of A.

In linguistic summaries of databases, we use fuzzy sets as models of vague linguistic expressions of quantities and properties of objects described by tuples (records), cf. Section 1.2.

Let $\mathscr{X}$, $\mathscr{Y}$ be universes of discourse, and $\mathscr{X} \times \mathscr{Y}$ be their (crisp) Cartesian product. Let A be a fuzzy set in $\mathscr{X}$. The *cylindric extension* of A to $\mathscr{X} \times \mathscr{Y}$ is the fuzzy set $\mathrm{ce}(A)$ in $\mathscr{X} \times \mathscr{Y}$, defined as:

$$\mathrm{ce}(A) = \left\{\left\langle \langle x,y\rangle, \mu_{\mathrm{ce}(A)}(x,y)\right\rangle : x \in \mathscr{X}, y \in \mathscr{Y}\right\} \tag{2}$$

where $\mu_{\mathrm{ce}(A)}(x,y) = \mu_A(x)$.

In general, if $\mathscr{X}_1, \ldots, \mathscr{X}_N$ are universes of discourse, and $\mathscr{X}_1 \times \ldots \times \mathscr{X}_N$ is their Cartesian product, and A is a fuzzy set in $\mathscr{X}_j$, $j \in \{1, \ldots, N\}$, the cylindric extension of A to $\mathscr{X}_1 \times \ldots \times \mathscr{X}_N$, is the fuzzy set $\mathrm{ce}(A)$ in $\mathscr{X}_1 \times \ldots \times \mathscr{X}_N$, defined as:

$$\mathrm{ce}(A) = \left\{\left\langle \langle x_1, \ldots, x_N\rangle, \mu_{\mathrm{ce}(A)}(x_1, \ldots, x_N)\right\rangle : x_1 \in \mathscr{X}_1, \ldots, x_N \in \mathscr{X}_N\right\} \tag{3}$$

where $\mu_{\mathrm{ce}(A)}(x_1, \ldots, x_N) = \mu_A(x_j)$.

1.2 Linguistic Summaries of Databases

We assume the following model of a database:

$$\mathscr{D} = \left\{\begin{array}{l} V_1(y_1),\ V_2(y_1),\ \ldots,\ V_N(y_1) \\ V_1(y_2),\ V_2(y_2),\ \ldots,\ V_N(y_2) \\ \qquad\qquad \ldots \\ V_1(y_m),\ V_2(y_m),\ \ldots,\ V_N(y_m) \end{array}\right\} = \left\{\begin{array}{c} d_1 \\ d_2 \\ \ldots \\ d_m \end{array}\right\} \tag{4}$$

where $\{y_1, \ldots, y_m\}$ is a set of objects, e.g. workers; $V_1, \ldots, V_N$ are attributes manifested by the objects, e.g. age, salary; $\mathscr{X}_1, \ldots, \mathscr{X}_N$ are the domains of $V_1, \ldots, V_N$, respectively, e.g. V_j is the age of workers and can take values from $\mathscr{X}_j = [20, 70]$. We denote a value of V_j for y_i as $V_j(y_i)$, $i \leq m$, $j \leq N$, e.g. V_j=Age, y_i='Smith' so $V_j(y_i) = 35 \in \mathscr{X}_j$. Thus, $d_i = \langle V_1(y_i), \ldots, V_N(y_i)\rangle$, $i = 1, 2, \ldots, m$, is the tuple describing y_i such that $d_i \in \mathscr{X}_1 \times \ldots \times \mathscr{X}_N$.

Linguistic summaries of databases are natural or quasi-natural language sentences, automatically built by computer software, and describing facts stored explicitly or implicitly in those databases [10, 11]:

$$Q\ P \text{ are/have } S_j\ [T] \tag{5}$$

where the symbols are interpreted as follows:

- Q is a determination of the amount (a quantity in agreement), i.e. linguistic quantifier, e.g. ABOUT HALF, SEVERAL [14]. The linguistic quantifiers are represented by fuzzy sets and may be *absolute*, e.g. ABOUT 5, LESS THAN 100, or *relative*, e.g. MOST, ABOUT HALF.
- P is the subject of the summary – the set of objects represented by tuples.
- S_j is a property of interest, the so-called *summarizer* represented by a fuzzy set in the domain of V_j, e.g. LOW TEMPERATURE in $\mathscr{X}_j = [-40, +60]$.
- T is a quality measure of the summary, *degree of truth*, expressed as a real number from the interval $[0,1]$, and interpreted as *level of confidence* for a given summary.

It is crucial for the linguistic summarization to know the algorithm computing the degree of truth T, strictly based on the Zadeh calculus of linguistically quantified statements [14]:

$$T\,(Q\,P \text{ are/have } S_j) = \mu_Q\left(\frac{r}{M}\right) \tag{6}$$

where $M = m$ if Q is relative, or $M = 1$ if Q is absolute, and

$$r = \sum\nolimits_{i=1}^{m} \mu_{S_j}(V_j(y_i)) \tag{7}$$

More frequently, we consider summaries with *compound summarizers*, i.e.

$$Q\,P \text{ are/have } \underbrace{S_1 \text{ AND/OR } S_2 \text{ AND/OR } \ldots \text{ AND/OR } S_N}_{S} [T] \tag{8}$$

see Section 3, and summaries with *qualifiers*, i.e.

$$Q\,P \text{ being/having } W \text{ are/have } S\,[T] \tag{9}$$

where W is *qualifier* – a property possessed by the objects and represented by a fuzzy set in $\mathscr{X}_W = \mathscr{X}_g$, $g \in \{1,\ldots,N\}$, see Section 4.

The gist of this paper is to show how degrees of truth, e.g. (6), and other quality measures of summaries can be evaluated using membership degrees of cylindric extensions of fuzzy sets.

2 Compound Linguistic Expressions Represented by Cylindric Extensions of Fuzzy Sets

To clarify the concept of using cylindric extensions of fuzzy sets in describing some values of linguistic summaries, we firstly need to show representation of compound linguistic expressions by two or more fuzzy sets in different universes of discourse[1]. In particular, we are interested in terms composed using connectives AND, OR. In

[1] In a broader context, one may discuss such expressions as labels of *linguistic variables* and their *compatibility levels*, cf. [13].

the Zadeh original approach [13], AND is modeled by the intersection of fuzzy sets associated to given terms, and OR – by the union of fuzzy sets [13].

We show here that the connectives AND, OR may be applied to the terms l_1, l_2 represented by fuzzy sets in different universes of discourse $\mathcal{X}_1$, $\mathcal{X}_2$, respectively. Let l_1, l_2 be modeled by fuzzy sets S_1, S_2 in $\mathcal{X}_1$, $\mathcal{X}_2$. Let y be an object described by real numbers $x_1 \in \mathcal{X}_1$, and $x_2 \in \mathcal{X}_2$. Terms composed of l_1, l_2 by AND, OR, are represented by the fuzzy sets S_{and}, S_{or} in $\mathcal{X}_1 \times \mathcal{X}_2$ which are the cylindric extensions of S_1 and S_2, both to $\mathcal{X}_1 \times \mathcal{X}_2$:

$$S_{and} = \mathrm{ce}(S_1) \cap \mathrm{ce}(S_2) \subseteq \mathcal{X}_1 \times \mathcal{X}_2 \tag{10}$$

$$S_{or} = \mathrm{ce}(S_1) \cup \mathrm{ce}(S_2) \subseteq \mathcal{X}_1 \times \mathcal{X}_2 \tag{11}$$

Equations (10) and (11) can be easily generalized to the case of $N \in \mathbb{N}$ linguistic terms $l_1, \ldots, l_N$ represented by fuzzy sets $S_1, \ldots, S_N$ in $\mathcal{X}_1, \ldots, \mathcal{X}_N$, respectively. Hence, the fuzzy sets S_{and}, S_{or} in $\mathcal{X}_1 \times \ldots \times \mathcal{X}_N$ are expressed by the intersection and sum, respectively, of the cylindric extensions $\mathrm{ce}(S_1)$, ..., $\mathrm{ce}(S_N)$ in $\mathcal{X}_1 \times \ldots \times \mathcal{X}_N$, cf. (3).

The generalizations are applied to linguistic summarization of databases, in particular, to build summarizers and qualifiers composed of several fuzzy sets.

3 Summaries with Compound Summarizers

In practice, we mostly deal with summaries which concern more than one attribute, e.g. *Some workers are tall and young*. The form is given by (8). Hence, the degree of truth of such a summary is evaluated as:

$$T\,(Q\;P\;\text{are/have}\;S_1\;\text{AND/OR}\;\ldots\;\text{AND/OR}\;S_N) = \mu_Q\left(\frac{r}{M}\right) \tag{12}$$

where $M = m$ for a relative Q, or $M = 1$ for an absolute Q.

The problem is to express the value r, since it is based on membership functions of several fuzzy sets $S_1, \ldots, S_N$. Hence, we propose to express the fuzzy set S (representing the summarizer) in terms of the unions and/or intersections of cylindric extensions of $S_1, \ldots, S_N$ to $\mathcal{X}_1 \times \ldots \times \mathcal{X}_N$:

$$\mu_S(d_i) = \mu_{\mathrm{ce}(S_1)}(d_i)\, t/s \ldots t/s\, \mu_{\mathrm{ce}(S_n)}(d_i),\ i = 1, 2, \ldots, m \tag{13}$$

where t is a t-norm representing the intersection, s is a t-conorm representing the union, and $\mu_{\mathrm{ce}(S_j)}(d_i) = \mu_{S_j}(V_j(y_i))$, $i = 1, \ldots, m$, $j = 1, \ldots, n$. Hence, r in (12) may be expressed as:

$$r = \sum_{i=1}^{m} \mu_S(d_i) \tag{14}$$

where $\mu_S \colon \mathcal{X}_1 \times \ldots \times \mathcal{X}_N \cap \mathcal{D} \to [0, 1]$.

4 Summaries with Qualifiers

It appears very useful to generate linguistic information on subsets of data which are distinguished by properties possessed by some, but not all, objects, e.g. one may be interested only in salaries of *young workers* out of all collected in a set. Therefore, Kacprzyk, Yager and Zadrożny [3] proposed the method of generation and evaluation of degrees of truth for summaries in the form of (9). Such summaries are potentially more interesting and descriptive. The degree of truth of such a summary is evaluated as:

$$T(Q\,P \text{ being/having } W \text{ are/have } S) = \mu_Q(r) \tag{15}$$

In this section, we use cylindric extensions of fuzzy sets representing W and S to express r and, in consequence, to evaluate T.

One may represent a qualifier W by several fuzzy sets $W_{g_1},\dots,W_{g_x}$, $g_1,\dots,g_x \in \{1,\dots,N\}$, in universes of discourse $\mathscr{X}_{g_1},\dots,\mathscr{X}_{g_x}$, respectively, in summaries e.g. *Many* YOUNG *or* INEXPERIENCED *workers have* LOW SALARIES. In particular, we represent the qualifier W as a fuzzy set in $\mathscr{X}_1 \times \dots \times \mathscr{X}_N$ with the membership function given as

$$\mu_W(d_i) = \mu_{\mathrm{ce}(W_{g_1})}(d_i)\,t/s \dots t/s\, \mu_{\mathrm{ce}(W_{g_x})}(d_i),\ i = 1,\dots,m \tag{16}$$

where $\mu_W\colon \mathscr{X}_1 \times \dots \times \mathscr{X}_N \cap \mathscr{D} \to [0,1]$, t,s are a t-norm and a t-conorm, respectively, representing the connectives AND, OR, respectively, and $\mu_{\mathrm{ce}(W_{g_k})}(d_i) = \mu_{W_{g_k}}(V_{g_k}(y_i))$, $i = 1,\dots,m$, $k = 1,\dots,x$.

Thus, to evaluate the degree of truth of a summary in the form given by (9), the following form of r is used:

$$r = \frac{\Sigma_{i=1}^{m}\left(\mu_S(d_i) \wedge \mu_W(d_i)\right)}{\Sigma_{i=1}^{m}\mu_W(d_i)} \tag{17}$$

where we still use (13) to determine the membership function $\mu_S(d_i)$, and the cofactor $\mu_W(d_i)$ in (17), means that only the tuples with the non-zero membership degrees to the property represented by the fuzzy set W – $\mu_W(d_i) > 0$ – are considered in the summary; other tuples are not considered.

5 Quality Measures

In this section, we intend to express some existing quality measures of linguistic summaries in terms of cylindric extensions. This new description can be very helpful in practical approaches, especially, in implementing and developing information systems based on linguistic summaries of databases, cf. [4, 7, 8], also in Java packages supporting fuzzy computations [9].

There exist many quality measures of linguistic summaries, defined in [2, 3, 5, 10]. In this section, we refer only to those for which an improved denotation using cylindric extensions of fuzzy sets is possible.

5.1 Defining Degree of Covering via Cylindric Extensions

The *degree of covering*, denoted as T_3, proposed e.g. in [3], is originally defined for summaries with qualifiers, see (9). The form of T_3 is proposed as

$$T_3 = \frac{\sum_{i=1}^{m} t_i}{\sum_{i=1}^{m} h_i} \tag{18}$$

where m is the number of tuples, and

$$t_i = \begin{cases} 1, \text{ if } \mu_S(d_i) > 0 \text{ and } \mu_W(d_i) > 0 \\ 0, \text{ otherwise} \end{cases}, \ i = 1, \ldots, m \tag{19}$$

and

$$h_i = \begin{cases} 1, \text{ if } \mu_W(d_i) > 0 \\ 0, \text{ otherwise} \end{cases}, \ i = 1, \ldots, m \tag{20}$$

where $\mu_S(d_i)$ is given by (13), $\mu_W(d_i)$ – by (16). Using the fuzzy sets which are defined as cylindric extensions of summarizers and qualifiers, T_3 can be expressed as:

$$T_3 = \frac{|\text{supp}(W \cap S \cap \mathscr{D})|}{|\text{supp}(W \cap \mathscr{D})|} \tag{21}$$

where $S \cap \mathscr{D}$ in $\mathscr{X}_1 \times \ldots \times \mathscr{X}_N$ is represented by the membership function $\mu_S \upharpoonright \mathscr{D}: \mathscr{D} \to [0,1]$, $(\mu_S \upharpoonright \mathscr{D})(d_i) = \mu_S(d_i)$, the fuzzy set $S \cap W \cap \mathscr{D}$ in $\mathscr{X}_1 \times \ldots \times \mathscr{X}_N$ is represented by $\mu_{S \cap W} \upharpoonright \mathscr{D}: \mathscr{D} \to [0,1]$, $(\mu_{S \cap W} \upharpoonright \mathscr{D})(d_i) = \mu_S(d_i)\, t\, \mu_W(d_i)$, $|\cdot|$ is the *Σcount* or *cardinality* of a fuzzy set[2], and supp$(\cdot)$ is the support of a fuzzy set[3].

It is also possible to redefine (18) with respect to summaries without qualifiers. The redefinition is based on the fact that we can express the form Q P are S, i.e. (8), as a particular case of the form: Q P being (in) $\mathscr{D}$ are S, i.e. based on (9). Therefore, we use W="being (in) $\mathscr{D}$" and it always refers to all tuples in the considered database $\mathscr{D}$. Hence, we rewrite (19) as

$$t_i^* = \begin{cases} 1, \text{ if } \mu_S(d_i) > 0 \\ 0, \text{ otherwise} \end{cases}, \ i = 1, \ldots, m \tag{22}$$

---

[2] The cardinality of a finite fuzzy set A in $\mathscr{X}$ is defined as $|A| = \Sigma\text{count}(A) = \sum_{x \in \mathscr{X}} \mu_A(x)$, cf. [1].

[3] $\text{supp}(A) = \{x \in \mathscr{X} : \mu_A(x) > 0\}$.

and (20) – as

$$h_i^* = 1 \ , \ i = 1, \ldots, m \tag{23}$$

because all the tuples belong to the set representing W, i.e to $\mathscr{D}$. Thus,

$$T_3^* = \frac{\sum_{i=1}^m t_i^*}{\sum_{i=1}^m h_i^*} = \frac{|\text{supp}(S \cap \mathscr{D})|}{m} \tag{24}$$

5.2 *Defining Degree of Appropriateness Using Cylindric Extensions*

Assume that a summarizer S in (8) or (9) is represented by a number of fuzzy sets $S_1, \ldots, S_N$, in $\mathscr{X}_1, \ldots, \mathscr{X}_N$, respectively, and for each of them the value r_j is computed as

$$r_j = \frac{\sum_{i=1}^m g_{ij}}{m} \ , \ j = 1, \ldots, N \tag{25}$$

where m is the number of tuples in $\mathscr{D}$, and

$$g_{ij} = \begin{cases} 1, \text{ if } \mu_{S_j}(V_j(y_i)) > 0 \\ 0, \text{ otherwise} \end{cases} \tag{26}$$

The values $r_1, \ldots, r_N$ are based on the function g_{ij} which counts the number of tuples having the property S_j and relates the sum of g_{ij} to m – the total number of tuples in $\mathscr{D}$. Hence, *degree of appropriateness*, T_4 is defined as:

$$T_4 = \left| \prod_{j=1}^n r_j - T_3 \right| \tag{27}$$

For this quality measure, we may use cylindric extensions of fuzzy sets in redefining the form of g_{ij} to:

$$g_{ij} = \xi_{\text{supp}(\text{ce}(S_j) \cap \mathscr{D})}(d_i) \tag{28}$$

which is based on the characteristic function, ξ, of the support[4] of the cylindric extension of the fuzzy set S_j to $\mathscr{X}_1 \times \ldots \times \mathscr{X}_N$ intersected with the set of tuples $\mathscr{D}$. In particular, the fuzzy set $\text{ce}(S_j) \cap \mathscr{D}$ is represented by the membership function $\mu_{\text{ce}(S_j)} \upharpoonright \mathscr{D}: \mathscr{D} \to [0,1]$, $(\mu_{\text{ce}(S_j)} \upharpoonright \mathscr{D})(d_i) = \mu_{ce(S_j)}(d_i)$, cf. [6].

Such forms of quality measures are much more handy in implementations which use object-oriented libraries dedicated for fuzzy computations, see [9].

6 Conclusions

The paper shows the use of cylindric extensions of fuzzy sets in a formal description of linguistic summaries of databases in the sense of Yager. Cylindric extensions of fuzzy sets make it possible to represent compound imprecise linguistic terms

[4] See Footnote 3.

associated to fuzzy sets in different universes of discourse. Hence, according to that, we rewrite some formulae evaluating quality measures for linguistic summaries, using these new descriptions. Besides, the used definition of the cylindric extension of a fuzzy set is fully compatible with the analogous definition for crisp (classic) sets, hence the former includes the latter as a special case.

In further work, it is possible to define cylindric extensions for other types of fuzzy sets, e.g. interval-valued or type-2 fuzzy sets, and use them in formal descriptions of extended methods of linguistic summarization of databases.

References

1. De Luca, A., Termini, S.: A definition of the non-probabilistic entropy in the setting of fuzzy sets theory. Information and Control 20, 301–312 (1972)
2. George, R., Srikanth, R.: Data summarization using genetic algorithms and fuzzy logic. In: Herrera, F., Verdegay, J.L. (eds.) Genetic Algorithms and Soft Computing, pp. 599–611. Physica–Verlag, Heidelberg (1996)
3. Kacprzyk, J., Yager, R.R., Zadrożny, S.: A fuzzy logic based approach to linguistic summaries of databases. International Journal of Applied Mathematics and Computer Sciences 10, 813–834 (2000)
4. Niewiadomski, A.: News generating via fuzzy summarization of databases. In: Wiedermann, J., Tel, G., Pokorný, J., Bieliková, M., Štuller, J. (eds.) SOFSEM 2006. LNCS, vol. 3831, pp. 419–429. Springer, Heidelberg (2006)
5. Niewiadomski, A.: Six new informativeness indices of data linguistic summaries. In: Szczepaniak, P.S., Węgrzyn-Wolska, K. (eds.) Advances in Intelligent Web Mastering, pp. 254–259. Springer, Heidelberg (2007)
6. Niewiadomski, A.: Methods for the Linguistic Summarization of Data: Applications of Fuzzy Sets and Their Extensions. Academic Publishing House EXIT (2008)
7. Niewiadomski, A.: A type-2 fuzzy approach to linguistic summarization of data. IEEE Transactions on Fuzzy Systems 16(1), 198–213 (2008)
8. Niewiadomski, A.: On Finity, Countability, Cardinalities, And Cylindric Extensions of Type-2 Fuzzy Sets in Linguistic Summarization of Databases. IEEE Transactions on Fuzzy Systems 18(3), 532–545 (2010)
9. Taborowski, L.: An intelligent computing library in java. Master's thesis, Institute of Computer Science, Technical University of Łódź, Poland (2005) (in Polish)
10. Yager, R.R.: A new approach to the summarization of data. Information Sciences 28, 69–86 (1982)
11. Yager, R.R., Ford, M., Canas, A.J.: On linguistic summaries of data. In: Piatetsky-Shapiro, G., Frawley, W.J. (eds.) Knowledge discovery in databases, pp. 347–363. AAAI Press, The MIT Press ((1991)
12. Zadeh, L.A.: Fuzzy sets. Information and Control 8, 338–353 (1965)
13. Zadeh, L.A.: The concept of linguistic variable and its application for approximate reasoning (I). Information Sciences 8, 199–249 (1975)
14. Zadeh, L.A.: A computational approach to fuzzy quantifiers in natural languages. Computers and Maths with Applications 9, 149–184 (1983)

Comparison of Selected Methods for Document Clustering

Radim Sevcik, Hana Rezankova, and Dusan Husek

Abstract. 17 cluster analysis techniques proposed for document clustering in terms of internal and external quality measures of clustering and computing time demands are compared. These are combinations of three basic methods (*direct, repeated bisection and agglomerative*) and five clustering criterion functions for solution assessment (*two intra-cluster, one inter-cluster, and two complex ones*); all implemented in the CLUTO software package. Furthermore, in the case of the agglomerative method we also applied a single linkage and complete linkage clustering as a criterion function. Collection 20 Newsgroups, a binary vector representation of e-mail messages, was used for comparing the methods. Experiments with document clustering have proved that, from the point of view of entropy and purity, the direct method provides the best results. As regards computing time, the repeated bisection (divisive) method has been the fastest.

Keywords: Web clustering, Cluster analysis, Textual documents, Web content classification, Newsgroups analysis, Vector model.

1 Introduction

The increasing size and dynamic content of the World Wide Web has posed the need of automated organization of data. Document clusters can provide a

Radim Sevcik · Hana Rezankova
University of Economics, Prague, nam. W. Churchilla 4,
13067 Praha 3, Czech Republic
e-mail: Sevcikrad@gmail.com, hana.rezankova@vse.cz

Dusan Husek
Institute of Computer Science, Academy of Sciences of the Czech Republic,
Pod vodarenskou vezi 2, 18207 Praha 8, Czech Republic
e-mail: dusan@cs.cas.cz

E. Mugellini et al. (Eds.): Advances in Intelligent Web Mastering - 3, AISC 86, pp. 101–110.
springerlink.com

structure for organizing large textual data for efficient browsing and searching. Document clustering has been traditionally investigated both as a means of improving the performance of search engines by pre-clustering the entire corpus, and as a post-retrieval document browsing technique [12, 1].

Search engines architecture is the part of information retrieval topic escribed in detail in [8, 5, 2, 7]. Here documents are usually modelled by high dimensional rectangular highly sparse document-word matrix with positive attribute values (type depends on data model) and significant amount of outliers.

The aim of clustering is either to create groups of similar objects or create a hierarchy of such groups. Any clustering technique relies on four concepts: model of data to be clustered, similarity measure, cluster model, and clustering algorithm that builds the clusters using the data model and the similarity measure, see [4].

A difficult task is an evaluation of the clustering quality. Various statistical approaches are used in this context, while in information retrieval we do this with the usual measures, such as precision and recall. Many different approaches for evaluation of the clustering quality are described, e.g. in [3].

The main contribution of this paper was the assessment of some available clustering methods from the point of view of their suitability for such document clustering. We used documents from 20 Newsgroups collection (e-mail messages collected from different newsgroups) solely in terms of their content as a testing corpora. After transforming documents into binary (Boolean) vector representation [10], we tested selected clustering methods, using different intra and inter quality functions available in software package CLUTO. For the case of clustering into 6 clusters we describe features of each cluster, mainly in terms of descriptive and discriminating words.

2 Applied Methods of Cluster Analysis

Three basic types of clustering methods (direct, repeated bisection and agglomerative) and five clustering criterion functions for solution assessment (two intra-cluster, one inter-cluster, and two complex ones) available in the CLUTO package, see [6], were applied. In addition, in the case of the agglomerative method we also applied a single linkage and complete linkage clustering as a criterion function.

We take a binary document-word matrix X as an input. Let us denote the number of documents by the letter n and the number of the words by the letter m. Thus, the matrix X is of the range n x m. A vector characterizing a certain document will be written as x_i, where $i = 1, 2, \ldots, n$.

2.1 Similarity Measure

The similarity of two documents was measured by a cosine function. When documents are characterized by binary vectors, we can express the numbers of combinations of the values 0 and 1 by the symbols in Table 1.

Table 1 Notation for the frequencies of combinations of 0 and 1 for two binary vectors

x_i	x_j	
	1	0
1	a	b
0	c	d

Then we can express the *cosine measure* as

$$s(x_i, x_j) = \sqrt{\frac{a}{a+b}\frac{a}{a+c}}. \tag{1}$$

This formula for binary data is called *Ochiai coefficient.* We used this measure as it is less demanding both as concerns the computing time as well as amount of the memory (instead of a multiplication of the number of documents and the number of words in the case of Pearson correlation coefficient only the number of non-zeros values are considered).

2.2 Clustering Criterion Functions

The documents were clustered to k clusters. Let us denote the hth cluster as C_h, $h = 1, 2, \ldots, k$ and the number of the documents in this cluster as n_h. Five clustering criterion functions were applied, see Table 2.

In addition to the criterion function mentioned above, single linkage and complete linkage approaches are used in agglomerative hierarchical cluster analysis.

2.3 Clustering Methods

Hierarchical cluster analysis (both agglomerative and divisive) and the so-called direct method were applied. In hierarchical cluster analysis, the similarity matrix is computed first.

The *agglomerative clustering* is represented by the *agglo* method in CLUTO. At the first stage of this method, individual documents are considered as clusters. Then, the most similar documents are joined into a cluster. The similarity matrix is recomputed and the process is repeated until k clusters are left. In the later stages, two smaller clusters are needed to join into one greater

Table 2 Formulas for criterion functions

Type	Symbol	Criterion function
Intra-cluster	I_1	$$\max \sum_{h=1}^{k} \left\{ \frac{1}{n_h} \sum_{x_i, x_j \in C_h} s(x_i, x_j) \right\} \quad (2)$$
Intra-cluster	I_2	$$\max \sum_{h=1}^{k} \sqrt{\sum_{x_i, x_j \in C_h} s(x_i, x_j)} \quad (3)$$
Inter-cluster	E_1	$$\min \sum_{h=1}^{k} n_h \frac{\sum_{x_i \in C_h, x_j \notin C_h} s(x_i, x_j)}{\sqrt{\sum_{x_i, x_j \in C_h} s(x_i, x_j)}} \quad (4)$$
Complex	H_1	$$\max I_1/E_1 \quad (5)$$
Complex	H_2	$$\max I_2/E_1 \quad (6)$$

cluster. For this purpose the similarity between two clusters is defined. In both linkage approaches, the similarities of all pairs of documents from two different clusters are computed. In the *single linkage approach (slink)*, for each pair of clusters the greatest value is considered as the similarity of the clusters. In the *complete linkage approach (clink)*, the smallest value is the similarity of the clusters. The most similar clusters are joined in both approaches.

Divisive clustering is represented by the *rb (repeated bisections)* method in CLUTO. At the first stage of this method, all the documents are in one cluster. This cluster is divided onto two smaller clusters. Then, one of these clusters is again divided into two clusters. The process is repeated until k clusters are created.

In the *direct* method, solution is found by simultaneously searching all k clusters. According to CLUTO manual, this approach is slower than clustering by the repeated bisection method. However, for small values of k (usually less than 10 – 20) it gives better clusters than via repeated bisections.

2.4 Quality Measures

Although a lot of internal and external measures for clustering quality evaluation have been proposed and described in the literature, see [3], here we use only the following measures: *average intra-cluster similarity* (ISim, average similarity between the objects of each cluster), the *standard deviation of the intra-cluster similarity* (ISdev), the *average inter-cluster similarity* (ESim, *the average similarity of the objects of each cluster and the rest of the objects*), and (ESdev *standard deviation of the inter-cluster similarity*). The best cluster is that for which the difference (ISim−ESim) is the greatest.

However, we usually need to compare different methods or different clustering, so one characteristic for the whole partitioning is useful. Therefore, we use the whole average internal similarity weighted by the numbers of documents in individual clusters and also weighted average inter-cluster similarity for the whole partitioning.

2.5 External Quality Measures

For the purpose of method evaluation in the case when the assignment of documents into groups is known in advance, we can compare the obtained results with the known assignment by entropy and purity.

Let us denote the number of clusters in the known partitioning (it can differ from the results obtained by clustering) as k' and the clusters in the known partitioning as $P_{h'}$, where $h' = 1, 2, \ldots, k'$. The *entropy of cluster* C_h is defined as

$$E(C_h) = -\frac{1}{\ln k'} \sum_{h'=1}^{k'} \frac{n_{hh'}}{n_h} \ln \frac{n_{hh'}}{n_h}, \tag{7}$$

where $n_{hh'}$ is the number of documents from the cluster C_h, which are also in the cluster $P_{h'}$. This measure takes on the values from 0 to 1.

The *entropy of the whole partitioning* is defined as

$$E(C) = \sum_{h=1}^{k} \frac{n_h}{n} E(C_h). \tag{8}$$

The *purity of cluster* C_h is defined as

$$P(C_h) = \frac{1}{n_h} \max_{h'=1,2,\ldots,k'} \left(\frac{n_{hh'}}{n} \right) \tag{9}$$

and the *purity of the whole partitioning* is defined as

$$P(C) = \sum_{h=1}^{h} \frac{n_h}{n} P(C_h). \tag{10}$$

This measure takes on the values from 0 to 1.

3 The 20 Newsgroups Data Set

This data set is a collection of approximately 20,000 messages gathered from 20 different newsgroups. Approximately one thousand messages from each of the twenty newsgroups were chosen at random and partitioned by a newsgroup name, see http://people.csail.mit.edu/jrennie/20Newsgroups. The documents are saved in the form of plain texts.

The list of newsgroups from which the messages were chosen is as follows: atheism (athe), graphics (grap), MS-Windows (os.m), IBM hardware (ibm.), Macintosh hardware (mac.), Windows XP (wind), for sale (fors), autos (auto), motorcycles (moto), baseball (base). hockey (hock), crypt (cryp), electronics (elec), medicine (med), space (spac), christianity (soc.), guns (guns), Middle East (mide), politics (poli), religion (reli). Some of the newsgroups are very closely related to each other and so we can distinguish six of the larger groups: computers (grap, os.m, ibm, mac., wind), recreation (auto, moto, base, hock), science (cryp, elec, med, spac), politics (guns, mide, poli), religion (athe, soc., reli) and miscellaneous (fors).

4 Steps of Prepared Data and Their Analyses

A typical process for preparing data from a collection of textual documents and their analyses is as follows: creation of vocabulary, preparation of the vectors representing individual documents and input matrix for the analyses, clustering of these vectors (i.e. documents), determination of the optimal number of clusters, and description of these clusters.

First, we created a vocabulary of the words used in the Rainbow system (http://www.cs.cmu.edu/ mccallum/bow), which is based on the Bow library, see [9]. For the purpose of this paper, i.e. a comparison of different clustering methods, approximately 100 documents from each group were randomly selected. For this small collection of 2,016 documents, a vocabulary containing 34,438 words was created.

On the basis of this collection and the corresponding vocabulary we prepared a set of binary vectors representing individual messages. The input matrix for the cluster analysis was prepared by means of the Awk system (http://www.manpagez.com/man/1/awk).

Then, several methods of cluster analysis were applied in the CLUTO system (http://glaros.dtc.umn.edu/gkhome/cluto/cluto/download), see [6]. This system was chosen due to its ability of large data sets clustering both in terms of the number of objects as well as the number of dimensions. A key feature of CLUTO clustering algorithms is an optimization process-seeking maximum or minimum of a criterion function, see [13]. This system provides criterion functions that make quality clustering solutions possible, especially for the case of high dimensional data sets characteristic for document data.

Moreover, methods are optimized for very sparse data sets, which is again the case of document data sets.

5 Comparison of Clustering Methods

We applied all the methods described in Sect. 2 for $k = 6$ and $k = 20$ clusters. In the first case, we expected sizes of clusters of about 500, 400, 400, 300, 300, and 100 documents (see Sect. 3). In the second case, it was about 100 documents for all clusters. We compared the methods by both internal and external quality measures and by the run time (on the computer Apple MacBook with Intel Core 2, 2.2 GHz, 4GB memory, Mac OS X 10.5.8).

We obtained the best results evaluated according to the run time by the *rb* method with the I_1 and I_2 criteria function. It was about 130 sec for 6 clusters and about 210 sec for 20 clusters in both cases. On the contrary, the most time-consuming methods were the *agglo* method with H_2 criterion for 6 clusters (almost 730 sec) and the *direct* methods with H_2 criterion for 20 clusters (1,025 sec).

As concerns cluster sizes, we obtained very bad results with the *agglo* method. In this particular case, we obtained one-element clusters almost for all the criterion functions with the exception of *clink*. In case of H_1 and H_2 criteria, there were five one-element clusters for clustering to six clusters, and four one-element clusters and one three-element cluster on case of the slink criterion. For 20 clusters the results were similar; only one big cluster was created with the use of H_1, H_2 criteria and *slink* criteria. On the other hand, the results obtained by the other methods proved to be satisfactory. For example, we obtained the sizes 523, 384, 367, 311, 310 and 121 documents for 6 clusters by the *direct* method while using the H_1 criterion. For 20 clusters we obtained the size of the smallest cluster 67 documents and the size of the largest cluster 204 ones by the *direct* method while using the E_1 criterion. Some other results differed only slightly.

From the point of view of entropy, the best results were obtained by the *direct* method with I_2 criterion both for 6 clusters (0.839) and for 20 clusters (0.708). The worst results were obtained by the *agglo* method with H_1, H_2, and *slink* criteria both for 6 clusters (0.997 for each of these methods) and for 20 clusters (0.99 for each of these methods).

In terms of purity, the best results were obtained by the *direct* method with I_2 criterion for 6 clusters (0.180) and by the direct method with H_1 criterion for 20 clusters (0.317). The worst results were obtained by the *agglo* method with H_1 and H_2 criteria both for 6 clusters (0.053 for each of both criteria) and for 20 clusters (0.06 for each of both criteria).

From the point of view of the whole average intra-similarity and whole average inter-similarity, the results were almost identical for the *direct* and *rb* methods (0.044 for the former and 0.023 for the latter were the best values for the six clusters and 0.062 and 0.025 were the best values for the twenty clusters). These results differed from those obtained by the *agglo* method.

6 Description of Obtained Clusters

For the description of created clusters by the typical words contained in them, we used the results obtained by the direct method with the I_2 criterion function for six clusters. Twenty origin groups included in the six created clusters (the numbers of the documents) are shown in Table 3.

Table 3 20 origin groups included in 6 created clusters (numbers of documents)

Groups	Clusters					
	1	2	3	4	5	6
athe	21	30	1	3	44	0
grap	17	11	0	17	9	46
os.m	15	23	1	4	0	57
ibm.	11	18	0	6	0	65
mac.	22	20	0	10	7	42
wind	8	15	1	11	7	59
fors	3	4	2	76	2	14
auto	23	31	11	11	21	4
moto	27	53	2	7	9	3
base	11	19	60	9	2	0
hock	8	7	77	5	4	0
cryp	25	37	0	4	33	2
elec	24	34	3	14	11	15
med	29	23	1	9	35	4
spac	18	40	3	3	30	7
soc.	15	17	2	5	62	1
guns	15	31	1	6	48	0
mide	8	30	1	0	62	0
poli	21	35	3	3	39	0
reli	24	35	1	1	40	0

We did not obtain clusters corresponding to the six groups described in Sect. 3 but we can see that the cluster 4 corresponds mainly to the group miscellaneous (fors – for sale) with 76 documents. The cluster 6 corresponds mainly to the group computers (grap, os.m, ibm, mac., wind) with 269 documents and the cluster 3 corresponds mainly to two small groups baseball (base) and hockey (hock) with 137 documents representing the topic sports. The remaining three clusters are created by a mixture of other groups with considerable representation of the groups moto with 53 documents and auto with 31 documents in the cluster 2 and considerable representation of the groups science (cryp, med, spac), politics (guns, mide, poli), and religion (athe, soc., reli) in the cluster 5.

We can obtain five the most descriptive and five the most discriminating words for each cluster in the CLUTO system. A descriptive word is chosen according to the benefit to average similarity between documents in the cluster. A discriminating word is chosen according to higher occurrence compared with other clusters. In our case, these words are representative particularly for the cluster 6 corresponding mainly to the group computers. Descriptive words are windows (4.3%), system (3.1%), work (3.0%), program (2.2%), and problem (1.9%). Discriminating words are writes (13.7%), article (12.4%), windows (3.9%), don (1.9%), and system (1.9%). For the cluster 3, the words game and team were denoted as both descriptive and discriminating. Other words are universal (writes, year, and article). For the cluster 4, the words mail, email, sale, address, and post were denoted as descriptive; the word sale was also denoted as discriminating.

In the remaining clusters, some other universal words, such as people, time, good, and make, were identified as descriptive or discriminating.

7 Conclusion

Our experiments with document clustering have shown that agglomerative clustering methods implemented in the CLUTO system are both very time-consuming and, what is more important, they do not give good results from different aspects. Exception is the case of the complete linkage approach. All methods are very sensitive to outliers and most clusters were of a very small size, often containing one element only.

Conversely, we have obtained relatively good results when using the direct and divisive clustering methods. The sizes of the clusters were adequate to the sizes of the groups from which documents were acquired. From the point of view of the whole average intra-similarity and whole average inter-similarity, the results were very similar for different criterion functions, and better than in the case of the agglomerative clustering. Methods evaluation obtained in this work is in accord with results of the work [11].

With regard to purity and entropy, we have obtained the best results when using the direct methods, mainly while applying the second intra-cluster criterion. Regarding the speed of calculation, the divisive method has proved to be the fastest.

We expected to achieve the best results in terms of purity and entropy for the case when number of clusters corresponds to the number of newsgroups (20), respectively to the number of clusters based on a human imposed classification (6). However, the results were rather bad. For example, the highest value of purity was only 0.317 for 20 clusters. These results can be influenced by some aspects, mainly by the quality of vocabulary and by binary vector representation instead of using weighted terms. Nevertheless, as the most descriptive and most discriminating words we have identified the groups computers, sports and for sale.

Acknowledgements. This work was supported by projects AV0Z10300504, GACR P202/10/0262, 205/09/1079, MSM6138439910, and IGA VSE F4/3/2010.

References

1. Andrews, N., Fox, E.: Recent Developments in Document Clustering. Tech. rep., Department of Computer Science, Virginia Tech. (2007)
2. Bouguila, N.: On multivariate binary data clustering and feature weighting. Computational Statistics and Data Analysis 54(1), 120–134 (2010)
3. Gan, G., Ma, C., Wu, J.: Data Clustering: Theory, Algorithms, and Applications. SIAM, Philadelphia (2007)
4. Husek, D., Pokorny, J., Rezankova, H., Snasel, V.: Data clustering: From documents to the web. In: Vakali, A., Pallis, G. (eds.) Web Data Management Practices: Emerging Techniques and Technologies, pp. 1–33. Idea Group Publishing, USA (2007)
5. Jiang, Z., Lu, C.: A latent semantic analysis based method of getting the category attribute of words. In: ICECT 2009: Proceedings of the 2009 International Conference on Electronic Computer Technology, pp. 141–146. IEEE Computer Society, Washington (2009), doi:10.1109/ICECT.2009.19
6. Karypis, G.: CLUTO: A Clustering Toolkit, Release 2.1.1. Tech. rep., University of Minnesota, Department of Computer Science, Minneapolis, MN (2003)
7. Li, T.: A unified view on clustering binary data. Machine Learning 62(3), 199–215 (2006), doi:10.1007/s10994-005-5316-9
8. Manning, C.D., Raghavan, P., Schütze, H.: Introduction to Information Retrieval, 1st edn. Cambridge University Press, Cambridge (2009)
9. McCallum, A.K.: Bow: A toolkit for statistical language modeling, text retrieval, classification and clustering (1996), `http://www.cs.cmu.edu/~mccallum/bow`
10. Sevcik, R.: Classification of Electronic Documents Using Cluster Analysis. Diploma thesis, University of Economics, Prague (2010)
11. Steinbach, M., Karypis, G., Kumar, V.: A Comparison of Document Clustering Techniques. Tech. rep., University of Minnesota, Department of Computer Science, Minneapolis, MN (2000)
12. Zamir, O., Etzioni, O.: Web document clustering: a feasibility demonstration. In: SIGIR 1998: Proceedings of the 21st Annual International ACM SIGIR Conference on Research and Development in Information Retrieval, pp. 46–54. ACM, New York (1998), doi: `http://doi.acm.org/10.1145/290941.290956`
13. Zhao, Y., Karypis, G.: Criterion Functions for Document Clustering. Tech. rep., University of Minnesota, Department of Computer Science (2003)

Speech Indexation in REPLAY

Samir Atitallah, Tobias Wunden, Maria Sokhn,
Elena Mugellini, and Omar Abou Khaled

Abstract. With the number of lecture recordings increasing constantly, technologies are needed to help the user find, access, and navigate through videos. This paper proposes a Sphinx-4-based [1] speech-to-text solution within the REPLAY framework to produce rich media and provide pin-point access.

Keywords: Speech-to-text, Sphinx4, REPLAY.

1 Introduction

The number of lectures and events being recorded has been growing constantly. REPLAY [2] was created to produce, manage, archive, and distribute audiovisual recordings on a large scale. An OCR-based indexation of the recorded slides is the basis of the videos being searchable in REPLAY. This kind of video indexation is an emergent topic in academia as these technologies provide the opportunity to navigate and show video on demand. The functionalities of visioning are constantly being improved, with video annotation being one of those features. Until today we only had the possibility to search videos using the title or the author of the video. Using this way of search, we lost almost all the information contained in the video.

Samir Atitallah
Ecole Polytechnique Federale de Lausanne, 1015, Lausanne
e-mail: samir.atitallah@epfl.ch

Tobias Wunden
ETH Zurich, ID-Multimedia Services, 8092, Zurich
e-mail: tobias.wunden@id.ethz.ch

Maria Sokhn · Elena Mugellini · Omar Abou Khaled
University of Applied Sciences of Fribourg,
Boulevard Perolles, 80, 1700, Fribourg
e-mail: maria.sokhn@hefr.ch, elena.mugellini@hefr.ch
omar.aboukhaled@hefr.ch

E. Mugellini et al. (Eds.): Advances in Intelligent Web Mastering - 3, AISC 86, pp. 111–119.
springerlink.com

Today the focus is on the content analysis of the videos. The main information of an academic video is contained in the slides and the audio part. In this project we focus on this last point. Analyzing the audio content of a video will give a lot of information which can be used to facilitate the user's searches. The purpose of this project is to analyze REPLAY and integrate a speech indexation functionality.

2 Context

REPLAY is a system which produces, manages, and distributes audiovisual recordings at ETH Zurich. The system integrates the recording of lectures and events, indexing, archiving and searching as the distribution of the content. It covers the complete life cycle of a multimedia production and incorporates media analysis functionalities with respect to the slides being captured. To record lectures at different locations at any time it is important that everything is automated. This is the essential point of REPLAY. The complete process, from the recording hardware to the distribution of the videos has to be done without any major manual intervention. This chapter will talk about the different phases of this automated mechanism. Every step is important and has to be optimized to ensure the quality and the rapidity of the system with large amounts of data.

- **Capture device: Playmobil** is the capture device of REPLAY. This is the first phase of the life cycle. It has been developed to capture high quality videos and VGA at the same time. This is essential to exploit the entire potential of the content. It is an automated recording system specifically designed for lectures and seminars.
- **Video Segmenter** is very important to REPLAY. The presentations are captured as videos which are called tracks. For the moment, REPLAY only indexes

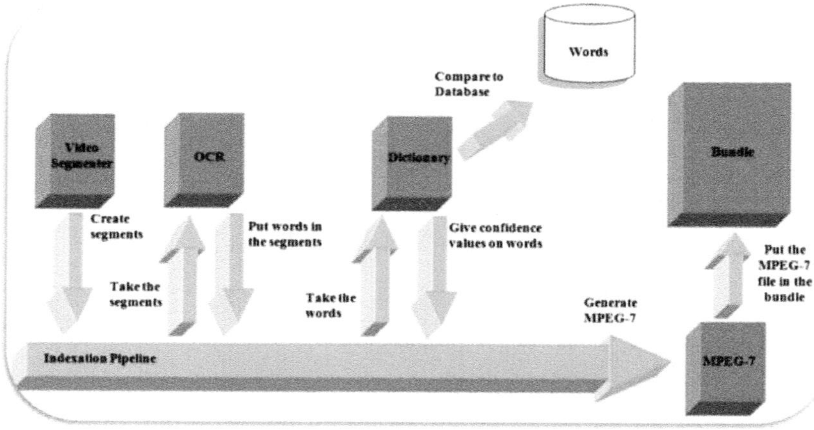

Fig. 1 REPLAY

presentation videos. It means that all the slides are captured as one video and there is no way to navigate into the different slides. For this, REPLAY integrates a Video Segmenter which can detect the transition between the slides and then cut the video into segments. There should be as many segments as slides. From now on, a segment is directly corresponding to a slide.

- **OCR and a relevance value** process analyzes each slide. The purpose of the OCR is to analyze the words which are contained in the slides and store them as isochronic metadata in the corresponding segment of the video. With the timeline it is easy to know on which segment the word is written.
- **Dictionary and confidence value** are there to help the OCR process. The OCR technology is very powerful in this context. However, this technology is not perfect. The content analysis of the slides using OCR is sometimes difficult. The analysis quality depends on the font size, the font colour, background images and other variables. It is not possible to control these because each lecturer brings his own presentation. Moreover, if the lecturer writes something on the slides using a tablet computer or moves the mouse on the text, the analysis quality can be deteriorated. Furthermore a dictionary step is implemented in REPLAY. Every word which was found by the OCR is compared to a dictionary to only use valid vocabulary.
- **MPEG-7** is used to store the metadata that are extracted from the OCR. The MPEG-7 file describes all the segments. This is the final step to get all the information centralized in a file. This file is used by the user interface to navigate into the video.

3 Speech-To-Text Plug-In

The integration of the speech-to-text plugin is the main part of this paper. REPLAY is already in production at ETH Zurich. Therefore it is essential that the speech-to- text plug-in doesnt interfere with the other plug-ins (Video Segmenter, OCR, and Dictionary). It should itself be a new functionality which integrates with the REPLAY system without modifying REPLAY. Therefore, the plug-in has been developed in a modular approach. It could be integrate in REPLAY or another system and work independently.

3.1 Speech Indexation Technologies

Voice recognition is the technology which will be introduce in this project. To analyze the audio content of the videos, it will be necessary to translate the spoken words into text. Once in text, we will have to integrate those words into REPLAY. The speech-to-text algorithm is just the first step of the mechanism. Speech-to-text is directly related to the voice recognition technology. It enables software to interpret natural human language. The principle is simple: a record of a few words delivered by a speaker on an audio track is interpreted as text. This technology uses methods from the domains of signal processing and artificial intelligence. A digitized

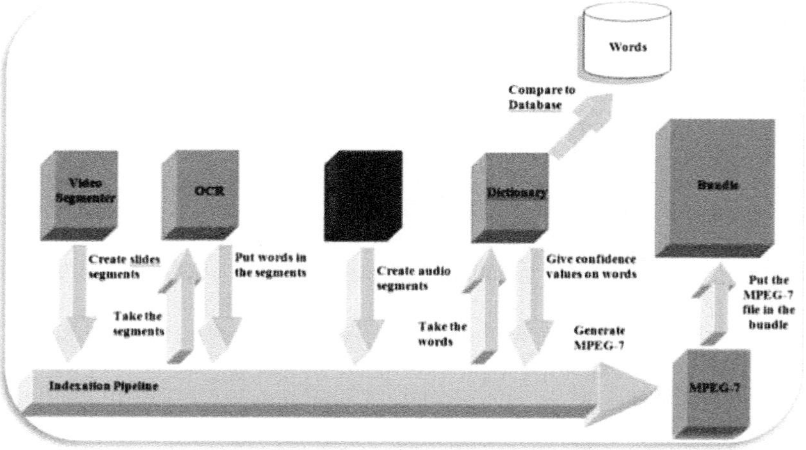

Fig. 2 REPLAY with speech indexation plug-in

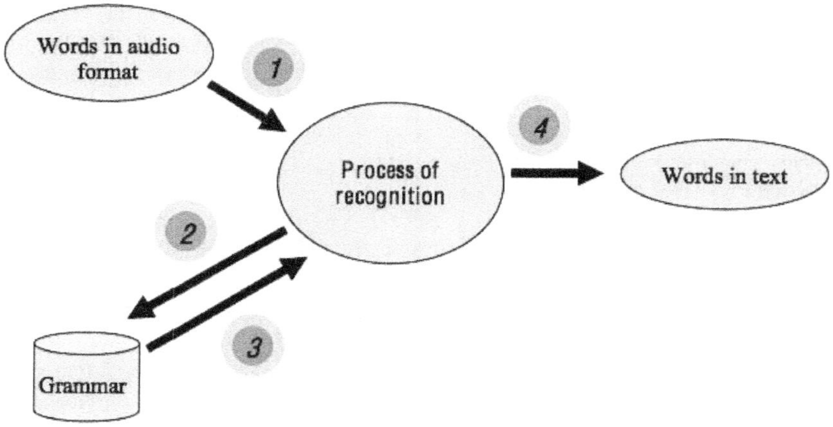

Fig. 3 Sphinx4 process

and recorded sentence is given to the programs voice recognition to translate them into words. There are several voice recognition software solutions, among the best known Crescendo, Dragon NaturallySpeaking and Sphinx4. The project REPLAY is based on a Java Platform. Robustness, stability and platform independence are essential to REPLAY. So the voice recognition library should be in Java, too. The Sphinx library seems to be the most appropriate for this use. It is the most recent library and it is the most flexible one. Sphinx4 has been developed by the University Carnegie Mellon, Sun Microsystems laboratory and Hewlett Packard. It is a very flexible and modular framework. Moreover the support is ideal and the framework and some examples of implementations are freely available as Open Source, which fits with the REPLAY licensing model and allows for future re-use in similar environments.

3.2 Storage of Audio Isochronic Metadata

MPEG-7 is a very complex but powerful technology used to store the metadata. It is implemented in REPLAY for several reasons. The first and maybe most important is that MPEG-7 is a standard. As said before, REPLAY has been implemented to be standardized as much as possible and to offer interoperability with other systems. It means that each plug-in or output of the system REPLAY can be used standalone and be understood without the context of REPLAY. Indeed, once the processing on the videos is done, the resulting metadata is stored. Then the exploitation of those data is infinite. It is important that the result is standardized to let every programmers or designer exploit this data without knowing anything about REPLAY. Even if MPEG-7 is a very complex technology and needs a lot of learning to become conversant with it, it is possible to do some basics things relatively easily. In addition, it offers the possibility of future expansions.The speech-to-text plug-in will use also this technology to store the metadata. This is the best way to integrate perfectly this plug-in and to benefit from the power of this technology.

3.3 Storage in REPLAY

Adding a speech-to-text plugin means there are now 2 different segmentations in REPLAY. The objective is to combine them in the most efficient, powerful and useful way. REPLAY provides the slide segmentation so that every user can access the part of the video that he is interested in. Since the slide content is the only point of reference for the user to explain to the system what he is searching for, the access points to each slide is good enough. With the speech-to-text recognition we will have another possibility of segmentation and access points to the audios content. The factors which are important are the efficiency of the speech-to-text implementation and the integration into REPLAY. In this case each technology will be stored separately. It means that the audio segments will be stored in a MPEG-7 file and the slide content in another, separate MPEG-7 file.

- **Advantages:**
  - Access to the audio words and the slides separately.
  - Different kinds of metadata are stored in different MPEG-7 files so it promotes the integration of further metadata sources (user comments, usage pattern etc.).
  - Choice of the sources of the metadata. It means that a developer use the specific MPEG-7 file if he wants for example treat only the audio metadata.

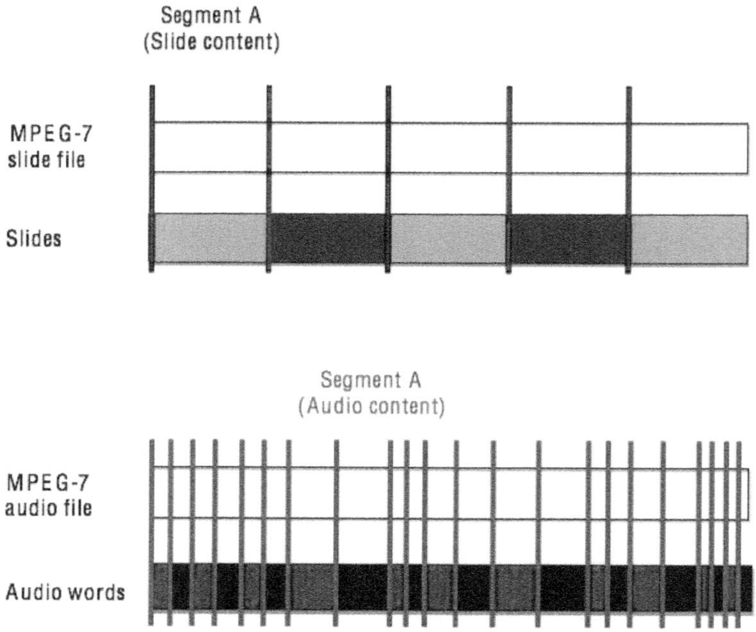

Fig. 4 Storage concept

- **Disavantages:**
  - The metadata is not centralised in one file

4 Audio Transcription Enhancement

Improving the efficiency of the speech-to-text algorithm is an important point. With the initial Sphinx4 technology, we get 78% correct words, so there is some room for improvements. Some optimization on the data and on the use of the library Sphinx4 can improve the quality for our use. The percentage of correct words in the transcripts is the key figure. This percentage expresses the correctness of the transcript content. This percentage determines if the transcript gives information which are correct. Once the REPLAY process is done, the transcript is the only result. It means that if there are a lot of wrong words in it, the speech indexation functionality will give wrong information. It is important that the search based on this audio transcript give correct results. Of course it is impossible for the moment to ensure that the user searches will always give a correct result. It is due to the speech-to-text algorithms which are not yet perfect. But we can improve this factor in 2 ways.

- **Optimisation by removing the rare words**
- **Optimisation by modifying the segments durations of the audio track**

The tests have been done on english videos with different pronunciation, audio quality and length.

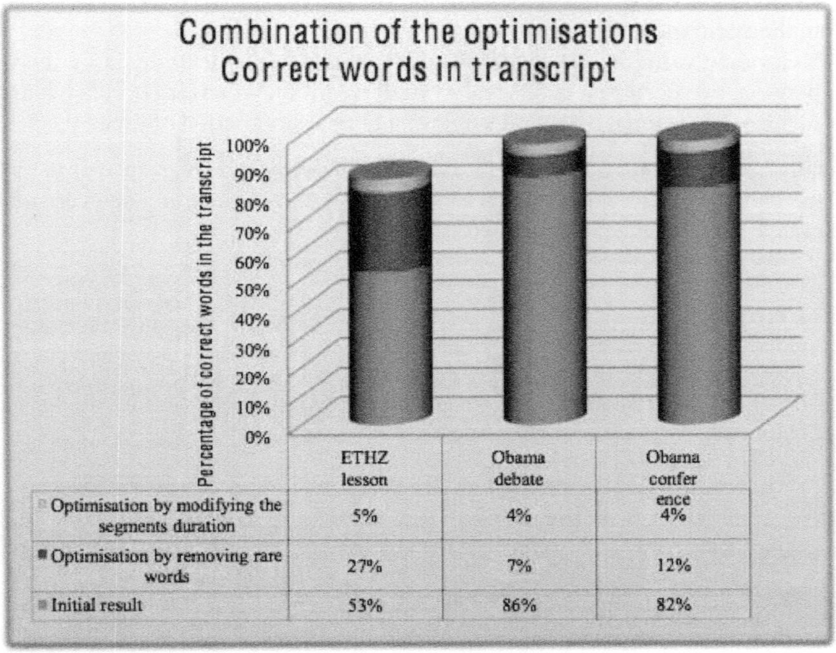

Fig. 5 Results with optimizations

5 Relevance Value: Algorithm Concepts

The relevance value defines the importance of a word. This relevance can be compared to the page rank in Google for example. When you search a word in Google, a list of websites are displayed. The list starts with the results which are very relevant.Defining relevance in speech-to-text plug-in is not so easy. In the OCR plug-in we can define it using the size and the location of a word on the slide. In the speech-to-text plug-in such measures are not available. In fact, we dont have a lot of information. We only know the words which were spoken and at what time in the video they were said. From that information only, a relevance value needs to be computed. To get a correct relevance value it is important to have some criteria. We will judge the relevance of a word on 3 criteria:

- **The relevance of the word in the entirety of the audio track.** To determine this relevance we will take into consideration the number of occurrences of each word in the entirety of the audio track. If a word is mentioned a lot of times, it is a relevant word within the video.
- **The relevance of the word in the context.** To determine this relevance we will take into consideration the number of occurrences of each word in an interval. If a word is mentioned a lot in a defined interval, it is a relevant word. This relevance can define if the word is relevant in the context of the respective current subject in the audio track.
- **Relevance value in the proximity.** This parameter is determined by the time between two occurrences of same word. It means that if a word occurs two times in a very short time, this word is relevant to this short period of time.

Finally each word has a relevance value which describes the relevance of the word in the entirety of the video, its relevance value in the context of a subject and the relevance value on the proximity words.

6 Prototype

To show the power of this plug-in, a prototype has been developed. This prototype is a Java GUI that contains a video media player and different tools to use the audio metadata.

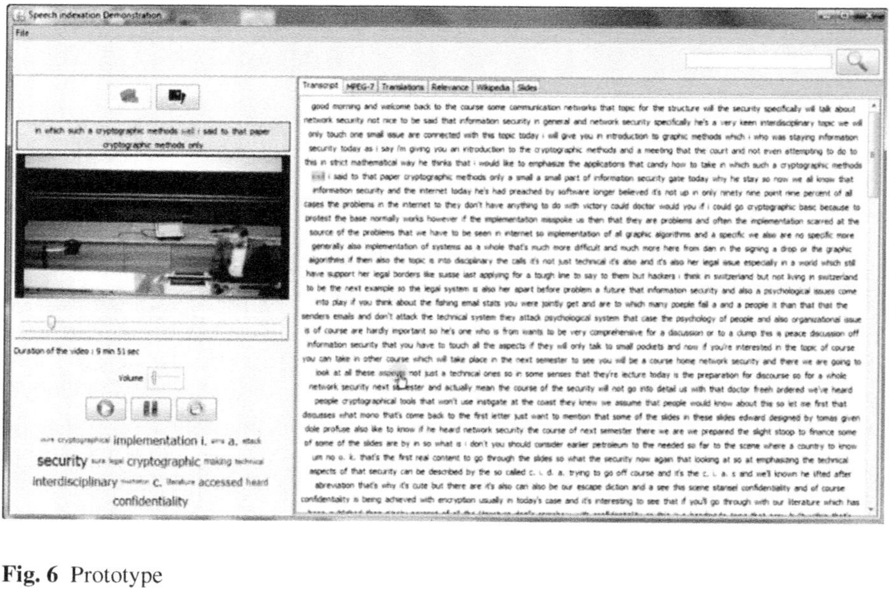

Fig. 6 Prototype

This "demonstration tool" contains the functionalities of the MIT Lecture browser [3] and a few others. It uses the MPEG-7 file of the audio segmentation in input and then offers:

- The monitoring of the transcript in real time (Current word is highlighted)
- Correction of the transcription
- Translation of the transcription
- Research of audio words by relevance
- Tag cloud of the most relevant words of the video

That prototype is accessible on http://speech.ethz.ch

7 Conclusion and Future Works

In this paper we presented a standardized speech-to-text plugin which offers opportunities for further analysis to developers. The prototype shows some interesting functionalities that could help students in their work. With the good performances of the Open Source speech-to-text algorithms we can imagine infinite ways of using the audio metadata. In addition, the plug-in is independent from REPLAY. It means that we will be able to integrate it to other system in the future. This plugin is an open door to a new kind of application.

References

1. Sphinx Homepage, `http://cmusphinx.sourceforge.net/sphinx4/`
2. REPLAY Homepage, `http://www.replay.ethz.ch/`
3. MIT Lecture Browser, `http://web.sls.csail.mit.edu/lectures/`

DegExt – A Language-Independent Graph-Based Keyphrase Extractor

Marina Litvak, Mark Last, Hen Aizenman,
Inbal Gobits, and Abraham Kandel

Abstract. In this paper, we introduce DegExt, a graph-based language-independent keyphrase extractor,which extends the keyword extraction method described in [6]. We compare DegExt with two state-of-the-art approaches to keyphrase extraction: GenEx [11] and TextRank [8].

Our experiments on a collection of benchmark summaries show that DegExt outperforms TextRank and GenEx in terms of precision and area under curve (AUC) for summaries of 15 keyphrases or more at the expense of a non-significant decrease of recall and F-measure. Moreover, DegExt surpasses both GenEx and TextRank in terms of implementation simplicity and computational complexity.

Keywords: Keyphrase extraction, summarization, text mining, graph-based document representation.

1 Introduction

Keyphrase extraction is defined as the automatic identification in a document of a set of terms that can be used to describe that document. Relevant extracted keyphrases, therefore, can be used to build an automatic index for a document collection, and they can be used for document

Marina Litvak · Mark Last · Hen Aizenman · Inbal Gobits
Department of Information System Engineering,
Ben-Gurion University of the Negev
Beer-Sheva 84105, Israel

Marina Litvak
Department of Software Engineering,
Sami Shamoon Academic College of Engineering
Beer-Sheva 84100, Israel

Abraham Kandel
Department of Computer Science and Engineering
University of South Florida
Tampa, FL 33620, USA

E. Mugellini et al. (Eds.): Advances in Intelligent Web Mastering - 3, AISC 86, pp. 121–130.
springerlink.com

representation in categorization or classification tasks. In addition, taken together, the keyphrases extracted from a certain document can function as an extractive summary of that document.

In this paper, we compare our graph-based extractor DegExt to two other approaches for keyphrase extraction—TextRank and GenEx—that are used in the extractive summarization of text documents. According to our problem statement, viable keyphrases are those that are listed in a gold standard document summary set. In the 1950s, Luhn [7] introduced a simple approach, based on using a frequency criterion, for selecting a document's keywords. Today, the state-of-the-art in keyword selection is represented by supervised learning methods, according to which a system is trained, based on lexical and syntactic features, to recognize keywords in a text.

The supervised learning approach for keyphrase extraction, first suggested by Turney [11], entails the combination of parameterized heuristic rules with a genetic algorithm (GA) to create the GenEx system, which automatically identifies keywords in a document. GenEx uses a GA to learn the best parameters for the extractor algorithm, with parameter values for the extractor as the population and the precision of the extractor as the fitness function. GenEx is based on the traditional vector-space model and is language-dependent: as a supervised algorithm, it cannot be adapted to new languages or domains without retraining it on every new type of data, and requires a high-quality corpus of annotated documents. Fortunately, Turney has shown that GenEx does generalize well to other domains.

Witten et al. [12] introduced Kea, another supervised approach using a Naïve Bayesian Decision rule with two features: tf-idf and the distance of the word from the beginning of the text.

Hulth [3] improved keyword extraction with a machine learning algorithm by adding linguistic knowledge (such as syntactic features) to the document representation instead of relying exclusively on statistics. The author showed that the results of any selection approach can be dramatically improved by extracting NP-chunks instead of n-grams and by adding the POS tag(s) assigned to each term as a feature.

All of the above approaches are supervised, and each uses a classic vector-space model for document representation. In contrast, Mihalcea and Tarau [8] introduced an unsupervised, graph-based keyphrase extractor called TextRank. TextRank utilizes a simple, syntactic graph-based representation for text documents, where nodes stand for unique non-stop words (more precisely, lexical units of a certain part of speech) connected by edges representing a *co-occurrence* relation, controlled by the distance between word occurrences: two vertices are connected if their corresponding words co-occur within a window of maximum N words, where $N \in [2, 10]$[1]. The main advantage of syntactic representation is its language-independency (given no syntactic filters) and simplicity—syntactic representation requires almost no

[1] Best results shown for $N = 2$.

language-specific linguistic processing. Mihalcea and Tarau [8] remark that vertices added to the graph can be restricted with syntactic filters, which select only lexical units of a certain part of speech. But for multilingual processing, TextRank can be run without syntactic filtering during the formative stages of document representation, and therefore, all words can be considered. Because we designed our experiments based on the results of Mihalcea and Tarau [8], we ran TextRank with a syntactic filter that focused only on nouns and adjectives as the best filter according to their results. Vertex importance within a graph was determined using PageRank, a graph-based ranking algorithm [1]. Formally, given the document graph $G(V, E)$, the score of a vertex V_i is defined by the formula:

$$S(V_i) = (1 - d) + d * \sum_{j \in In(V_i)} \frac{S(V_j)}{Out(V_j)}$$

where $In(V_i)$ is the set of vertices that connect with vertex V_i, $Out(V_i)$ is the set of vertices that vertex V_i connects with (successors), and d is a damping factor that integrates, into the model, the probability of jumping from a given vertex to another random vertex in the graph. Usually, the damping factor is set to 0.85, which is the value used in TextRank implementation. The top ranked vertices are extracted as the keywords. Post-processing, in which adjacent document keywords are collapsed into multi-word keyphrases, is then performed.

Recent papers explore World Wide Web knowledge like Wikipedia in order to model documents as a semantic network. Grineva et al. [4] introduce title-community approach that uses the Girvan-Newman algorithm to cluster phrases into communities and selects those phrases in the communities containing the title phrases as key phrases. Li et al. [5] propose a novel semi-supervised keyphrase extraction approach by computing the phrase importance in the semantic network, through which the influence of title phrases is propagated to the other phrases iteratively.

2 DegExt — Degree-Based Extractor

Like TextRank, DegExt is an unsupervised, graph-based, cross-lingual keyphrase extractor. DegExt uses graph representation based on the *simple* graph-based syntactic representation of text and web documents defined in [10]), which enhances the traditional vector-space model by taking into account some structural document features. The *simple* graph representation holds unlabeled edges representing order-relationship between the words represented by nodes. The stemming and stopword removal operations of basic text preprocessing are performed before graph building[2]. A single vertex is created for each distinct word, even if the word appears more than once in the text. Thus, each vertex label in the graph is unique.

[2] This part may be skipped for multilingual processing unless appropriate stemmers and stopword lists are provided for different languages.

Unlike original *simple* representation, where only a specified number of most frequent terms are added into graph, we don't have any constraints on the number of graph nodes. In our system, filtering of nodes may be specified by configurable parameters using the absolute number of nodes or ratio threshold. However, in our experiments we did not limit the number of nodes at all, in order to avoid the dependency on additional parameter.

Edges represent order-relationships between two terms: there is a directed edge from A to B if an A's term immediately precedes a B's term in any sentence of the document. However, in the event that sentence-terminating punctuation marks (periods, question marks, and exclamation points) are present between the two words, an edge is not created. In order to adapt the graph representation to multi-word keyphrase extraction, we label each edge by the IDs of sentences that contain both words in the specified order. This definition of graph edges is slightly different from co-occurrence relations used in [8] for building undirected document graphs holding unlabeled edges, where the order of word occurrence is ignored and the size of the co-occurrence window varies between 2 and 10.

Figure 1 shows a sample text (enumerated sentences) and its graph representation respectively.

0	Hurricane Gilbert Heads Toward Dominican Coast.
1	Hurricane Gilbert swept toward the Dominican Republic Sunday, and the Civil Defense alerted its heavily populated south coast to prepare for high winds, heavy rains and high seas.
2	The storm was approaching from the southeast with sustained winds of 75 mph gusting to 92 mph.

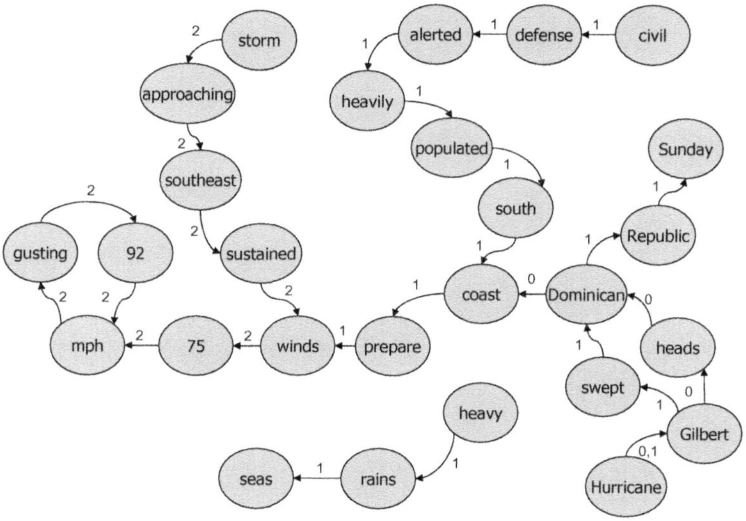

Fig. 1 Text and its graph representation

This representation can be extended to many different variations like a semantic graph where nodes stand for concepts and edges represent semantic relations between them or a more detailed syntactic graph where edges and nodes are labeled by significant information like frequency, location, similarity, distance, etc. For HTML documents, edges may be labeled by section ids: *title*, which contains the text related to the document's title and any provided keywords (meta-data) and *text*, which comprises any of the readable text in the document.

The syntactic graph-based representations were shown by Schenker et al. [9] to perform better than the classical vector-space model on several clustering and classification tasks. We chose to use the *simple* representation as a basis in this work because it is relatively cheap in terms of processing time and memory resources while having provided nearly the best results for the two above text mining tasks.

The most connected nodes in a document graph are assumed by DegExt to represent the keywords. When document representation is complete, every node is ranked by the extent of its connectedness with the other nodes, and the top ranked nodes are then extracted. Intuitively, the most connected nodes, i.e., the top ranked nodes, represent the most salient words. According to the above representation, words that appear in many sentences that diverge contextually (i.e., the surrounding words change from sentence to sentence) will be represented by strongly connected nodes. This intuition was approved by experimental results in [6], where different number of HITS iterations were tried and the best result was achieved after the first iteration, ranking nodes by their degree. Thus, we showed that applying ranking algorithms to document graphs (using the *simple*-based representation) does not improve the extractor performance on English corpora, but even makes it worse.

In order to identify keyphrases (as a sequences of adjacent keywords), DegExt scans the document graph during postprocessing marking all selected potential keywords in the graph, and sequences of adjacent keywords (up to 3) having the same label on edges between them are combined into a multi-word keyphrase. The posprocessing proposed by [8] is also applicable. The final rank for each phrase is calculated as an average between the ranks for each of its words. N (specified by the user) top-ranked phrases are extracted as keyphrases.

Since the DegExt algorithm is involved with constructing document representation, sorting graph nodes by degree, and identifying keyphrases, it has much lower computational complexity than TextRank, which needs additional time $O(c(|E| + |V|))$ to run PageRank. Here c is the number of iterations needed to converge, $|E|$ is the number of edges, and $|V|$ is the number of nodes (words) in a document graph. Representation building in both algorithms has the same computational complexity. When DegExt is used for document representation tasks without syntactic filtering, it is absolutely language-independent.

3 Experimental Results

All experiments were performed on the benchmark collection of summarized news articles provided by the 2002 Document Understanding Conference (DUC) [2]. This collection contains 533 English texts, each with an average of 2-3 abstracts (gold standard abstracts) per document. To evaluate the extraction results, we ran the keyphrase extractors on the DUC document collection and compared the extracted keyphrases against the gold standard abstracts. We used common metrics such as precision, recall, F-measure and AUC (Area Under Curve). Selected keyphrases that appeared in at least one abstract for a given document were considered *true positives*, selected keyphrases that did not appear in any abstract were *false positives*, keyphrases that were not selected and that did appear in the abstracts were considered *false negatives*, and keyphrases that were not selected and that did not appear in abstracts were considered *true negatives*. Since all evaluated extractors output keyphrases along with single words, we created an inverted index of phrases from the gold standard abstracts for comparison purposes. The average size[3] of the syntactic graphs compiled from the phrases extracted from these texts was 212, and it varied from 66 to 944.

Statistics on Six Decades of Oscar With PM-Oscar Nominations Bjt

The motion picture industry's most coveted award, Oscar, was created 60 years ago and 1,816 of the statuettes have been produced so far.
Weighing 8{ pounds and standing 13{ inches tall, Oscar was created by Metro-Goldwyn-Mayer studios art director Cedric Gibbons, who went on to win 11 of the trophies.
Oscar, manufactured by the R.S. Owens Co., Chicago, is made of Britannia metal, copper plate, nickel plate and gold plate.
From 1942 to 1944, the trophy was made of plaster, but winners were later presented with the real thing.
According to the Academy of Motion Pictures Arts and Sciences, the only engraving mistake was in 1938 when the best actor trophy given to Spencer Tracy for ``Boy's Town" read: ``Best Actor: Dick Tracy."
The Academy holds all the rights on the statue and ``reserves the right to buy back an Oscar before someone takes it to a pawn shop," said Academy spokesman Bob Werden.
The most-nominated film was ``All About Eve" in 1950.
It got 14 nominations.
``Ben-Hur" in 1959 was the most-awarded film with 11, and Walt Disney was the most-awarded person with 32.

Fig. 2 Text document from DUC 2002 collection

[3] We define the size of a graph as the number of its vertices.

plate, academy motion,
actor, engraving,
art director

Fig. 3 Extraction results for $N = 5$ with TextRank

oscar, plate, most-
awarded film, rights,
best actor trophy

Fig. 4 Extraction results for $N = 5$ with DegExt

oscar,
academy,
plate, film, actor

Fig. 5 Extraction results for $N = 5$ with GenEx

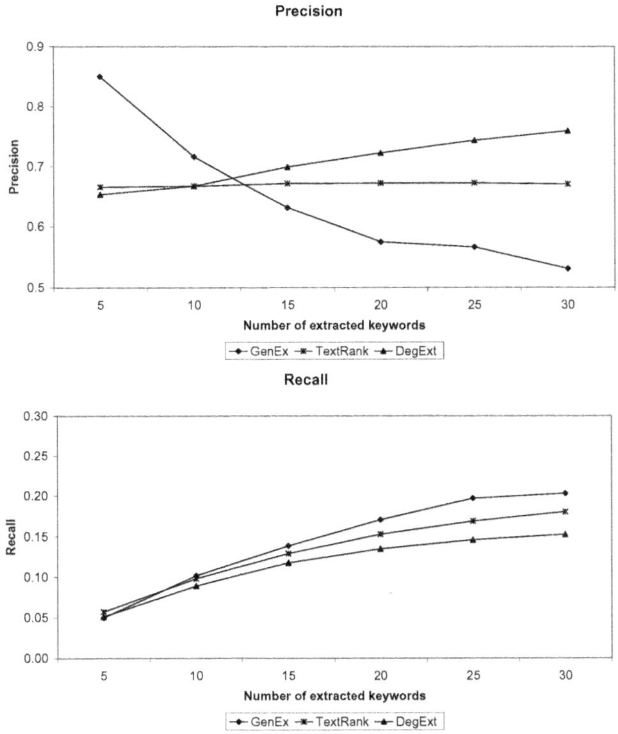

Fig. 6 Evaluation results for GenEx, TextRank, DegExt and six models respectively (5 - 30 keyphrases)

Figures 3, 4, and 5 present five resulting keyphrases for TextRank, DegExt, and GenEx, respectively, in one of the English documents entitled "Statistics on Six Decades of Oscar With PM-Oscar Nominations Bjt" (in boldface). Figure 7 demonstrates the precision, recall, F-measure, and AUC values for each of the methods evaluated on the English corpus. We considered six summary models distinguished by the number of top ranked phrases—from 5 to 30, granularity of 5—each extracts.

GenEx had the highest precision and AUC values (up to 10 and 15, respectively) for "small" models. The best precision value for GenEx can be explained by using the precision as a fitness function in the GA. Also, GenEx had the highest, but not significantly distinguishable, recall and F-measure results. Since GenEx not always succeeds to extract as many keypharses as required, its Precision decreases with the number of needed keyphrases.

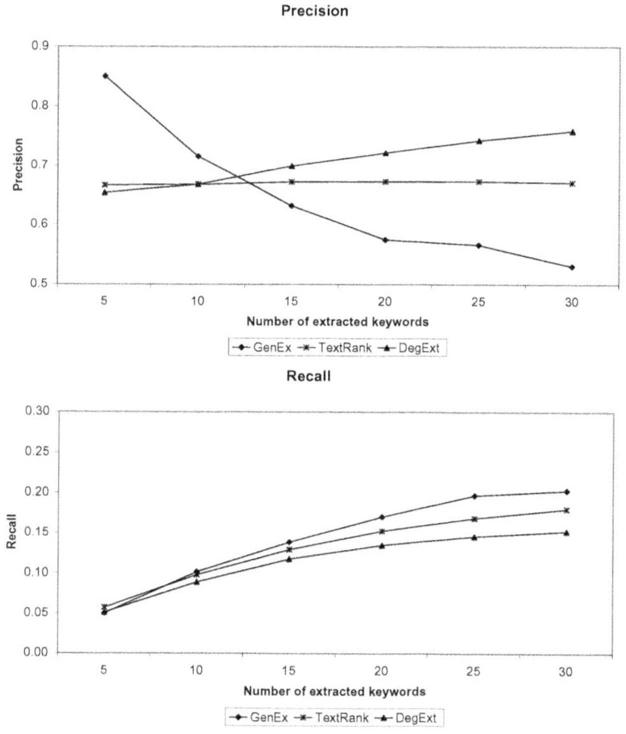

Fig. 7 Evaluation results for GenEx, TextRank, DegExt and six models respectively (5 - 30 keyphrases)

DegExt exhibited the best values for precision and AUC for the "large" models that extract greater numbers of required keyphrases (above 10 and 15, respectively), but those high values were obtained at the expense of relatively lower recall and F-Measure values. For example, for 20 required extracted keyphrases, the F-measure of DegExt was approximately 15% and 10% lower than the highest (GenEx) and the second highest (TextRank) values, respectively. Moreover, DegExt precision was approximately 30% and 15% better than the lowest (GenEx) and the second lowest (TextRank) values, respectively.

As an unsupervised algorithm, DegExt does not require time for training, and its computation time is equal to the time required for it to build document representation. Assuming efficient implementation, DegExt has linear computational complexity relative to the total number of words in a document ($O(n)$) with node sorting taking logarithmic time.

4 Conclusions and Future Work

In this paper we introduced DegExt – a graph-based keyphrase extractor for the extractive summarization of text documents. We compared DegExt with two approaches to keyphrase extraction: GenEx and TextRank.

Our empirical results suggest that the supervised GenEx approach has the best precision (a finding that can be explained by using precision as a fitness function of the GA) and AUC for small numbers of extracted keyphrases. However, the major disadvantages of this approach are a long training time and language dependency. Inspite of its good performance, GenEx is a supervised learning method with an expensive computational complexity for the training phase [11], it should be retrained in order to perform well on different types of documents and multiple languages[4].

When there is no high-quality training set of significant size and a large number of keyphrases (above 15) is needed, we recommend using the unsupervised method based on node degree ranking—DegExt—which provides the best precision and AUC values for large numbers of of keyphrases. According to our experimental results, we can extract up to 30 phrases with an average precision above 75%, an average recall above 15%, and an F-measure above 24%.

Performance of the DegExt approach surpasses those of the other evaluated approaches—GenEx and TextRank—in terms of implementation simplicity and computational complexity. A major advantage that both TextRank and DegExt have over GenEx is their language-independence.

In our future research, we intend to evaluate our method on additional languages to demonstrate the cross-linguality of our approach. Also, other graph representations of documents may be evaluated.

[4] However, the version of GenEx tool that we used cannot be applied to any language except English.

References

1. Brin, S., Page, L.: The anatomy of a large-scale hypertextual web search engine. Computer Networks and ISDN Systems 30, 107–117 (1998)
2. DUC (2002), Document understanding Conference, http://duc.nist.gov
3. Hulth, A.: Improved automatic keyword extraction given more linguistic knowledge. In: Proceedings of the 2003 Conference on Empirical Methods in Natural Language Processing, Japan (2003)
4. Grineva, M., Grinev, M., Lizorkin, D.: Extracting key terms from noisy and multitheme documents. In: Proceedings of the 18th International Conference on World Wide Web, Madrid, Spain, pp. 661–670 (2009)
5. Li, D., Li, S., Li, W., Wang, W., Qu, W.: A Semi-Supervised Key Phrase Extraction Approach: Learning from Title Phrases through a Document Semantic Network. In: Proceedings of the ACL 2010 Conference Short Papers, Uppsala, Sweden, pp. 296–300 (2010)
6. Litvak, M., Last, M.: Graph-based keyword extraction for single-document summarization. In: Proceedings of the Workshop on Multi-source Multilingual Information Extraction and Summarization, pp. 17–24 (2008)
7. Luhn, H.P.: The automatic creation of literature abstracts. IBM Journal of Research and Development 2, 159–165 (1958)
8. Mihalcea, R., Tarau, P.: Textrank – bringing order into texts. In: Proceedings of the Conference on Empirical Methods in Natural Language Processing, Barcelona, Spain (2004)
9. Schenker, A., Bunke, H., Last, M., Kandel, A.: Classification of web documents using graph matching. International Journal of Pattern Recognition and Artificial Intelligence 18, 475–496 (2004)
10. Schenker, A., Bunke, H., Last, M., Kandel, A.: Graph-theoretic techniques for web content mining. World Scientific Pub. Co. Inc., Singapore (2005)
11. Turney, P.D.: Learning algorithms for keyphrase extraction. Information Retrieval 2, 303–336 (2000)
12. Witten, I.H., Paynter, G.W., Frank, E., Gutwin, C., Nevill-Manning, C.G.: Kea: practical automatic keyphrase extraction. In: Proceedings of the Fourth ACM Conference on Digital Libraries, Berkeley, California, USA, pp. 254–255 (1999)

Verifying Authenticity in Interactive Behaviors of Semantic Web Services

Xiaolie Ye and Lejian Liao

Abstract. Due to the importance of security within Semantic Web services, every interaction protocol embedded in dynamic behaviors of composed services should be formally modeled and verified for the satisfaction of some security requirement, such as the compliance of authentication, authorization and privacy policy. Our purpose is to model interactive behaviors and verifying security properties in the ontology-base semantic layer. Towards the aim, we present an OWL-based Past Linear Temporal Logic (Past-LTL) to describe temporal properties within interactions of Semantic Web services and refine some algorithms to reduce the validity of an OWL-based Past-LTL formula into the entailment relationship in OWL. With the help of the action theory on describing dynamic aspect of Semantic Web services, we propose an approach to transform the verification of the authenticity in interactive behaviors of Semantic Web services into the validity of the OWL-based Past-LTL formula corresponding.

Keywords: Interaction Protocol, Temporal Logic, Authenticity.

1 Introduction

Due to the importance of security within Semantic Web services, every interaction protocol embedded in dynamic behaviors of composed services should be formally modeled and verified for the satisfaction of some security requirement, such as the compliance of authentication, authorization and privacy policy. But, no matter the

Xiaolie Ye and Lejian Liao
Beijing Institute of Technology, 5 South Zhongguancun Street, Haidian District, Beijing, China
e-mail: yexiaolie@hotmail.com
Liaolj@bit.edu.cn

E. Mugellini et al. (Eds.): Advances in Intelligent Web Mastering - 3, AISC 86, pp. 131–141.
springerlink.com

approach to SOAP[1]-based cryptographic protocols or a traditional formal verification is not a sufficient solution. From interactive behaviors of semantic Web services, it is possible to separate a SOAP-based cryptographic protocol, through mapping the parameters in service profile to the XML-based messages in the service ground w.r.t OWL-S[2], and verify the security properties using the approach mentioned above. However, since a SOAP-base encapsulation is just constituted on XML syntax, ontology semantics of parameters will lost in the ground layer. Perhaps, it causes ignoring some potential flaw in semantics. Moreover, traditional formalisms to verify the security are based on the assumption of closed world (the unknown is false in default). In the context of semantic Web services, the functionalities and non-functionalities, such as service of qualities and security policies, are defined in an ontology language, such as OWL. But, as the basis of OWL, a family of Description Logics (DLs)[1] is based on the open world assumption.

So, the key points to solve the problems are how to describe interactive behaviors, static information, and the security properties required in the ontology-based semantic layer, and how to validate those properties by a reasonable reduction method. Since the situation calculus [12] and GOLOG [11] are applied to formalize the dynamic aspects of Web services and to describe their composition, Franz Baader, etc[2], integrate action theory with DLs to model actions associated with functionalities of Web services and present an approach to reasoning for their composition. The major problem in these approaches is that reasoning in incomplete knowledge of the world, such as whether a service, composed with a sequence of invocations to atomic services, is executable. In other words, an atomic service is defined as an action that has a pre-condition and post-condition (a set of effects), described using description logic assertions, and the current state is the incompletely knowledge of the world as a set of such assertions (a so-called ABox). Then, a composed service is simply defined as a sequence of actions (atomic services). In result, the *composability* of semantic Web services is reduced into the consistency of a knowledge base in DLs. So that, an appreciated approach to the key points can be established on the action-based formalism of modeling and reasoning for dynamic aspects of Semantic Web services mentioned above.

For simplicity, we only consider how to validate the authenticity in an interaction protocol within Semantic Web services. The intuitional explanation of authenticity is that an event e authenticates an agent a such that e can occurs only if a previous message send by a [13]. So that, one agent should become sure of the identity of the other. Usually, the authenticity is described as a temporal property. Moreover, we only concern this context that an authentication protocol, embedded in interactive behaviors of atomic Semantic Web services, is formalized as a sequence of actions described in the fragment of OWL. Then, we simplify the action-based approach to model interactive behaviors within Semantic Web services and propose a method to reason for the validity of temporal properties along a sequence of changes caused by interactions of those. The contribution is listed as follows:

---

[1] Simple Object Access Protocol.

[2] Ontology Web Language for Services to provide the building blocks for encoding rich semantic service descriptions, http://www.w3.org/Submission/OWL-S.

1. As a variant temporal logic, a OWL-based Past-Linear Temporal Logic is proposed to formalize temporal properties along a *path* (state changes caused by a sequence of the execution of atomic services) w.r.t the *ALCQIO*[3] fragment of OWL-DL;
2. We reduce the authenticity into a Past-LTL formula within an interaction protocol, and give a demonstration to verify it using our approach.

In the paper, the rest is organized as follows: Section 2 presents Past-LTL and a skeleton of the reduction approach. In Section 3, we express the authenticity in an interacting protocol as a Past-LTL formula. Finally, we present the related works in Section 4 and give some conclusions in Section 5.

2 OWL Ontology Based Past-LTL and Reasoning

2.1 The *ALCQIO fragment* of OWL

In DLs, *ALCQIO* allows for these constructors including Negation, Conjunction, Disjunction, Existential and Universal restrictions, Qualified number restriction, Inverse role, and Nominal. In this paper, we focus on the fragment of OWL-DL corresponding to *ALCQIO* with the abbreviation *fragment*, where *Classes* are inductively defined starting with a set N_A of *named classes* or *atomic classes*, a set N_r of *named object properties*, and (possibly) a set N_I of *named individuals*. In this paper, we abbreviate the OWL syntax as follows: a named class and class is denoted with A and C (possibly with a subscription), an object property r (possibly with a subscription), and an individual a (possibly with a subscription) in *fragment*. Furthermore, for presenting the semantics of a *fragment*, we take the direct model-theoretic semantics in OWL 2 [4] and simplify those. Namely, given a 4-tuple $< \Delta_I, \cdot^C, \cdot^{OP}, \cdot^I >$ as an interpretation I , in which Δ_I is a non-empty set of object domains and $\cdot^C$ is a mapping to assign a subset C^C of Δ_I to each class C, $\cdot^{OP}$ is a mapping to assign a subset r^{OP} of $\Delta_I \times \Delta_I$ to each object property r, and $\cdot^I$ is a mapping to assign a subset a^I of Δ_I to each individual a. And also, we call an interpretation I as a model such that I satisfies an ontology O if all of conditions resulted from each axiom in O are satisfied by I, noted with $I \models O$(I Possibly with subscription to indicate a time point.). One of the most important relationships between two ontologies is the entailment. Namely, let O_1 and O_2 be tow ontologies, O_1 entails O_1, noted with $O_1 \models O_2$, such that for every interpretation I, $I \models O_1$ implies $I \models O_2$. Likewise, let φ be an *fragment* axiom, then $O_1 \models O_2$ if and only if for every $\varphi \in O_2$, $O_1 \models \varphi$. Respecting *fragment*, apparently, every class expression has only a simple object property. So that, the inference, e.g. the ontology entailment, is decidable.

[3] The prototypical Description Logics Attributive Concept Language with Complements is the basis of many expressive Description Logics.

[4] `http://www.w3.org/TR/2009/REC-owl2-direct-semantics-20091027/`

2.2 Conceptualizing Assertion Change

For formalizing an interaction protocol, we present a concept *assertion change(AC)* that describes a change from a source *fragment* ontology into another.

Definition 1. (Assertion Change) The *assertion change* is a set α of atomic assertions (e.g. class assertion, object property assertion, and negative object property assertion axioms) that describe a change on a *fragment* ontology.

Supposed that we follow the constant domain assumption, (namely, all interpretations share the same set of individuals within a common domain, let I_1 and I_2 be the different models for a *Fragment* ontology, then $domain((\cdot^I)_{I_1}) = domain((\cdot^I)_{I_2}).$), we present Def.2 to express a transition from one model to another.

Definition 2. (AC-Labelled Transition) Let α be AC that puts effect on a model I_1 w.r.t a *fragment* ontology O, and products another model I_2 w.r.t O. If the following is hold for every atomic class A and object property r w.r.t O, we call that an $AC-labelled\ transition(ALT)$ noted with $I_1 \rightarrow_\alpha I_2$.

$$A^{I_2} = (A^{I_1} \cup \{a^{I_1} | a : A \in \alpha\} \cdot \backslash \{b^{I_1} | b : \neg A \in \alpha\}) \tag{1}$$

$$r^{I_2} = (r^{I_1} \cup \{(a^{I_1}, b^{I_1}) | (a,b) : r \in \alpha\} \backslash \cdot \{(c^{I_1}, d^{I_1}) | (c,d) : \neg r \in \alpha\}) \tag{2}$$

And, according to Def.2 above, we propose another definition *ALT path* as the following:

Definition 3. (ALT Path) The *ALT Path(path)* is defined as a sequence ρ of $AC-labelled\ transitions$, noted with follows: (Note that a subscription i indicates the order in ρ.)

$$\rho = \begin{cases} I_0, i = 0, \\ I_0 \rightarrow_{\alpha_1} I_1 \rightarrow_{\alpha_2} \cdots \rightarrow_{\alpha_i} I_i, i \geq 0. \end{cases} \tag{3}$$

Given a *path* ρ and $0 \leq i \leq \#\rho$, let $\alpha = f_{AC}(\rho, i)$ be a function to obtain an AC α at the position i in ρ.

Furthermore, through combining some past temporal operators with assertions in a *fragment* ontology, we could obtain Past-LTL formulae as defined in Def.4. And, each of the past temporal operators, such as Y(Yesterday), and U^-(until in history), only appears in front of an assertion.

Definition 4. (Past-LTL Formula) Given a *fragment* ontology O, a Past-LTL formula is inductively defined as follows:

1. For every $\varphi \in O$, φ is a Past-LTL formula.
2. If either of φ and ψ is a Past-LTL formula, then $\varphi \wedge \psi, \varphi \vee \psi$, $Y\varphi$, and $\varphi U^- \psi$, are also Past-LTL formulae.

Then, as to a Past-LTL formula, back along a *path*, the validity is defined as follows:

Definition 5. (Validity along a *path*) Given a Past-LTL formula φ and a *path* ρ from an initial *fragment* ontology O, the validity of a Past-LTL formula φ, back from a time point i along ρ, noted with $(\rho,i) \models \varphi$, is inductively defined as follows:
$(\rho,i) \models \varphi$, iff for the interpretation I_i, $I_i \models \varphi$ and φ is an assertion axiom;
$(\rho,i) \models \varphi \wedge \psi$, iff $(\rho,i) \models \varphi$ and $(\rho,i) \models \psi$;
$(\rho,i) \models \varphi \vee \psi$, iff $(\rho,i) \models \varphi$ or $(\rho,i) \models \psi$;
$(\rho,i) \models \neg\varphi$, iff $(\rho,i) \not\models \varphi$;
$(\rho,i) \models Y\varphi$, iff $(\rho,i-1) \models \varphi, i > 0$; Otherwise, false;
$(\rho,i) \models \varphi U^- \psi$, iff $\exists k\ 0 \leq k \leq i, (\rho,k) \models \psi$, implies $\forall j\ k \leq j \leq i$, $(\rho,j) \models \varphi$;

Each interpretation in a *path* can be one of the possible worlds as similar as in first order logics. So, a Past-LTL formula can be used to express some temporal properties for a sequence of behaviors that change the possible world.

2.3 Reducing

There, we refine the reducing method in [3] and propose an OWL-based approach for validating a Past-LTL formula φ along a *path* ρ from a *fragment* ontology O. Supposed that each Past-LTL formula φ has been NNF, given a time point i in a *path*, the reducing skeleton is as the following steps:

1. Due to the meaning of *ALT* in Def.2, we can define some equivalent class axioms Γ_i^{red} to conceptualize the minimization of changes on individuals, classes, and axioms at each transition in ρ from O.
2. In a *path* ρ, the changes on the named objects will be guaranteed by a set Λ_i^{red} of the reduced assertions that consists of the initial assertions, the *AC* assertions in ρ , and the preserving assertions;
3. The Past-LTL formula φ will be transformed into a set ∂ of sets of assertion axioms by a set of reducing tableaux rules;
4. Finally, φ is valid such that $\exists \Lambda_i^{\varphi}, \Lambda_i^{\varphi} \in \partial, \Gamma_i^{red} \sqcup \Lambda_i^{red} \models \Lambda_i^{\varphi}$;

Let Ξ be a triple $< O, \rho, \varphi >$ as an input that consists of a *fragment* ontology O , a *path* ρ, and a Past-LTL formula φ. Let *Sub* be a set of all class expressions occurring in Ξ, and Λ a set of assertion axioms in O. And, A, r, and T is an atomic class expression (an named class), an object property or a negative object property expression, and a class expression for an *AC*, respectively. Syntactically, either of names and expressions is possibly with a subscript that indicates a time point in a *path*; and, it is also with a superscript that indicates which constructor the conceptualization is related to .

Following Law of Inertia in Action theory [12], changes between tow interpretations should be little as possible while still satisfying all post-conditions. Intuitively, the minimization of changes on named elements can be described in a direct way through Λ_i^{red}, while the minimization of changes on unnamed elements is achieved through a suitable encoding of T in Γ_i^{red}. As mentioned in [2], since the interpretation of a defined class is uniquely determined by the interpretation of an atomic class and role names, it is sufficient to impose this minimization of change condition on named classes and roles.

Given an input Ξ, Γ_i^N is defined as a set that contains a single equivalence class axiom for a named class N and a conjunction of all nominal classes within Ξ as shown in (4). Furthermore, T_k^A in (5) stands for the equivalent class to the interception of the atomic class A after the kth transition. As shown in the right side of the formula (5), the *unionOf* $\sqcup$ constructor connects tow parts: the first, expressing named elements and the second, expressing the unnamed elements. For same reason, the equivalent class to the interception of each named class, such as $T_k^{C\sqcup D}$, $T_k^{C\sqcap D}$, $T_k^{\neg C}$, $T_k^{\exists r.C}$, $T_k^{\forall r.C}$,$T_k^{\geq nr.C}$, and $T_k^{\leq nr.C}$ in Sub can be inductively defined by the semantics of constructors. Finally, we can get a set Γ_k^{Sub} of equivalent class axioms. In addition, as shown in (6), besides the axioms reduced from nominal and AC, Γ_i^{red} also contains others axioms reduced from initial equivalent class axioms in O, and the object property domain and range axioms w.r.t Ξ.

$$\Gamma_i^N = \{N \equiv \bigsqcup_{0 \leq k \leq i} \{a_k\}\} \tag{4}$$

$$T_k^A \equiv (N \sqcap A_k) \sqcup (\neg N \sqcap A_0), A \in Sub, \tag{5}$$

$$\begin{aligned} \Gamma_i^{red} = \Gamma_i^N \sqcup (\bigsqcup_{0 \leq k \leq i} & (\Gamma_k^{Sub} \sqcup \{T_k^A \equiv T_k^E | (A \equiv E) \in O\} \\ & \sqcup \{Domain(r_k, T_k^C) | Domain(r, C) \in O\} \\ & \sqcup \{Range(r_k, T_k^C) | Range(r, C) \in O\})) \end{aligned} \tag{6}$$

In this paper, we only discuss the case adding new assertions(possibly a negative object property). And also, with the addition of domain and range axioms, our approach avoids producing too many reduced assertions to affect the availability.

Given an input Ξ, with reducing each of class expressions, we also need to reduce the assertions related. For all assertions w.r.t Ξ, at a time point i or after a transition, each class expression occurring within a class assertion or object property axiom should be replaced with the reduced one. Let φ be a class assertion $a : C$, or an object property assertion $a, b : r$, or a negative object property assertion $a, b : \neg r$, then the reduced one, as shown in (7), be φ_i, or $a, b : r_i$, or $a, b : \neg r_i$ at time point i, respectively.

$$\varphi_i = \begin{cases} a : T_i^C), & \text{if } \varphi = a : C \\ (a, b) : r_i, & \text{if } \varphi = a, b : r \\ (a, b) : \neg r_i, & \text{if } \varphi = a, b : \neg r \end{cases} \tag{7}$$

Given Λ as an initial set of all assertions w.r.t Ξ, Λ^{ini} is a set of the results through (8).

$$\Lambda^{init} = \{\varphi_0 | \varphi \in \Lambda\} \tag{8}$$

So, the set Λ_i^{red} of reduced assertions is defined inductively as follows(9 - 14):

$$\Lambda_0^A = \Lambda_0^{\neg A} = \Lambda_0^r = \Lambda_0^{\neg r} = \Lambda^{init} \tag{9}$$

$$\begin{aligned} \Lambda_k^A =& \Lambda_{k-1}^A \cup \{a : A_{k-1} \rightarrow A_k | a \in N_I, \text{ and } a : A_{k-1} \in \Lambda_{k-1}^A\} \\ & \cup \{a : A_k | a : A \in f_{AC}(\rho,k)\}, \text{ if } 1 \leq k \leq i. \end{aligned} \tag{10}$$

$$\begin{aligned} \Lambda_k^{\neg A} =& \Lambda_{k-1}^{\neg A} \cup \{a : \neg A_{k-1} \rightarrow \neg A_k | a \in N_I, \text{ and} \\ & a : \neg A_{k-1} \in \Lambda_{k-1}^{\neg A}\} \cup \{a : \neg A_k | a : \neg A \in f_{AC}(\rho,k)\}, \text{ if } 1 \leq k \leq i. \end{aligned} \tag{11}$$

$$\begin{aligned} \Lambda_k^r =& \Lambda_{k-1}^r \cup \{a : (\exists r_{k-1}\{b\} \rightarrow \exists r_k.\{b\}) | a,b \in N_I, \text{ and } (a,b : r_{k-1}) \in \Lambda_{k-1}^r\} \\ & \cup \{a,b : r_k | a,b : r \in f_{AC}(\rho,k)\}, \text{ if } 1 \leq k \leq i. \end{aligned} \tag{12}$$

$$\begin{aligned} \Lambda_k^{\neg r} =& \Lambda_{k-1}^{\neg r} \cup \{a : (\forall r_{k-1}\neg\{b\} \rightarrow \forall r_k.\neg\{b\}) | a,b \in N_I, \text{ and} \\ & (a,b : \neg r_{k-1}) \in \Lambda_{k-1}^{\neg r}\} \cup \{a,b : \neg r_k | a,b : \neg r \in f_{AC}(\rho,k)\}, \\ & \text{if } 1 \leq k \leq i. \end{aligned} \tag{13}$$

$$\Lambda_{red} = \Lambda_i^A \cup \Lambda_i^{\neg A} \cup \Lambda_i^r \cup \Lambda_i^{\neg r} \tag{14}$$

As to a Past-LTL formula φ, through a tableaux approach, φ is unfolded into a set Λ_i^{φ} of assertions w.r.t Ξ. In the tableaux rules (15-18), ∂ is a set of sets of Past-LTL formulae with an initial status $\partial = \{\{\varphi_i\}\}, \varphi_0 = \varphi$. In $\vee Rule$, the set β' and β'' is defined in (19-20) And, in $U^{-}Rule$, Ω_k is defined as (21).

$$\frac{\Lambda \in \partial \wedge (\varphi \wedge \phi)_i \in \Lambda}{\Lambda := (\Lambda \backslash \{(\varphi \wedge \phi)_i\}) \cup \{\varphi_i, \phi_i\}} \wedge Rule \tag{15}$$

$$\frac{\Lambda \in \partial \wedge (\varphi \vee \phi)_i \in \Lambda}{\partial := (\partial \backslash \{\Lambda\}) \cup \{\Omega', \Omega''\}} \vee Rule \tag{16}$$

$$\frac{\Lambda \in \partial \wedge (Y\varphi)_i \in \Lambda,\ 0 \leq i \leq \#\rho}{\Lambda := (\Lambda \backslash \{(Y\varphi)_i\}) \cup \{\varphi_{i-1}\}} YRule \tag{17}$$

$$\frac{\Lambda \in \partial \wedge (\varphi U^{-} \phi)_i \in \Lambda,\ 0 \leq i \leq \#\rho}{\partial := (\partial \backslash \{\Lambda\}) \cup \{\Omega_i, \Omega_{i-1}, \ldots, \Omega_0\}} U^{-}Rule \tag{18}$$

$$\Omega' = (\Lambda \backslash \{(\varphi \vee \phi)_i\}) \cup \{\varphi_i\} \tag{19}$$

$$\Omega'' = (\Lambda \backslash \{(\varphi \vee \phi)_i\}) \cup \{\phi_i\} \tag{20}$$

$$\Omega_k = (\Lambda \backslash \{(\varphi U^{-} \phi)_i\}) \cup \{\varphi_i, \varphi_{i-1}, \ldots, \phi_{i-k},\ 0 \leq k \leq i\} \tag{21}$$

Finally, The rules above are applied exhaustively on ∂ to get rid of any temporal operator in φ. And, we can take a set Λ_i^{φ} from ∂, which should be as candidates to check the entailment relationship with the reduced ontology $\Gamma_i^{red} \cup \Lambda_i^{red}$.

Theorem 1. *Let Ξ be an input as a triple (O, ρ, φ) and O, ρ, and φ, be an initial framgment ontology, a path, and a Past-LTL formula, respectively. Then, given a time point i in ρ, through (4-5), Γ_i^{red} is obtained as a set of reduced defining axioms; and through (9-14), Λ_i^{red} as a set of reduced asserting axioms. Likewise, through (15-18) w.r.t Ξ, ∂ is obtained as a set of sets of assertions for φ. Then, $(\rho, i) \models \varphi$ if and only if $\Gamma_i^{red} \cup \Lambda_i^{red}$ is consistent and $\exists \Lambda_i^{\varphi} \in \partial$, $\Gamma_i^{red} \cup \Lambda_i^{red} \models \Lambda_i^{\varphi}$.*

In summary, the validity of a Past-LTL formula w.r.t the input is transformed into the problem checking the entailment relationship between tow OWL ontologies (tow sets of axioms).

3 Authenticity in Past-LTL

The authenticity is an essential property for interacting protocols always as a temporal property. And, with supporting from [13, 4, 8], we represent the property related to the authenticity in Past-LTL. Usually, the correspondence or non-injection in the authentication is expressed in a temporal logic based on such events, e.g. beginning an initial request, ending an initial request, etc[8]. So, we follow it but express them as a Past-LTL formula in a *fragment*, so as to conceptualize each event occurring in an authentication procedure. There are six events C^{BI}, C^{RI}, C^{EI}, C^{BR}, C^{RR}, and C^{ER} in Tab.3 with other classes and object properties related. And, as to (22), we can obtain the concrete event assertion only if each E is substituted by one of C^{BI},

Table 1 Conceptualizing Authentication Events in *Fragment*

Fragment	Abbreviation	Statement
:Role	C^R	a participant.
:Session	C^S	a session.
:Nonce	C^N	random value.
:Begin(End)Init	C^{BI}	begin or end in an initiator.
:RunInit	C^{RI}	run an authentication in an initiator.
:Begin(End)Response	C^{BR}	begin or end in a responser.
:RunResponse	C^{RR}	run an authentication in a responser.
:who	r^w	range over participants.
:session	r^s	range over sessions.
:nonce	r^n	range over random value.
:parter	r^p	range over participants.

C^{RI}, C^{EI}, C^{BR}, C^{RR}, and C^{ER}.

$$e:E(i,p,s,n) \doteq ((e:E)\wedge(e,i:r^{w})\wedge(p:r^{p})\cdot\wedge(s:r^{s})\wedge(e,n:r^{n})\cdot \\ \wedge(i:C^{R})\wedge(p:C^{R})\cdot\wedge(s:C^{S})\wedge(n:C^{N}) \tag{22}$$

Given a collection of individual variables a, b, s, and n and tow event individuals e_0, e_1, a Past-LTL formula (23) with a universal quantification $\forall(a:C^{R},b:C^{R},s:C^{S},n:C^{N})$, represents the authenticity within an interacting protocol. (Note that $\odot$ is the 'Once' operator and $\odot\varphi$ is as $(\top U^{-}\varphi)$, and $\phi\rightarrow\varphi$ is as $\neg\phi\vee\varphi$.)

$$\forall(a:C^{R},b:C^{R},s:C^{S},n:C^{N})\cdot(e_0:C^{ER}(b,a,s,n)\rightarrow \\ \odot(e_1:C^{RI}(a,b,s,n))) \tag{23}$$

As a result, given an abbreviated Past-LTL formula $\forall v\delta$ as (23) and an *input* Ξ, the authenticity is hold in ρ such that for every instantiation of the individual variables a, b, s, and n, $\rho\models\delta$ w.r.t Ξ. (Note that $\rho\models$ is the abbreviation of $(\rho,\#\rho)\models$.).

4 Related Works

Karthikeyan Bhargavan and Cedric Fournet et al [5], propose a specification language TulaFale for writing machine-checkable descriptions of SOAP-based security protocols and their properties, which can be complied into the applied pi calculus, and be verified using Blanchet's resolution-based protocol verifier [6]. Moreover, E. Kleiner and A.W. Roscoe [10] propose a method for mapping interacting messages to abstract symbols in the style of Dolev-Yao, and Casper notation and formally analyze WS-Security and WS-SecureConversation. While these approaches mainly consider how to specify, model and verify security of SOAP-based interactions between Web services, our approach focus on modeling and reasoning for security properties, such as authenticity, in an ontology-based semantic layer. That makes our approach more suitable for open environment.

For the security aspects of composing services, Barbara Carminati et al put efforts on security constraint-based Web services composition [7]. Moreover, Lalana Kagal et al present some ontology of policy language and a distributed solution for policy management to enhance the traditional identification and access control framework so as to realize the dynamic and non-center management[9]. Although these works discuss the enforcement of ontology-based security policies within a dynamic context, such as the composition of services, the reasoning mechanism is limited to check the satisfaction of static security properties since the absence of semantics of interactions (actions).

5 Conclusions and Future Works

With the help of the action theory and OWL, we can formalize a sequence of *state changes* caused by the invocations of atomic Semantic Web services and check the salification of temporal properties under incomplete knowledge of world. In particular, Authenticity as a concrete temporal property, has been expressed as an OWL-base Past-LTL formula. According to the approach, the validity of temporal properties in an interaction protocol has been reduced into obtaining an entailment relationship, namely, detecting whether a set of the axioms reduced from a *path* entails the one unfolded from the Past-LTL formula for the temporal properties. For more clarity, some algorithms has been proposed for reducing a *path* and unfolding a Past-LTL based on the result of the former. As a concrete application, we have represented the mechanism marking events in a *fragemnt* for checking the non-injective agreement and reduce the authenticating procedure into a *path*. As a result, verifying the authenticity in an interacting protocol has been reduced into the validity of a Past-LTL formula in a *path*.

Since the result is archived on simplifying the action theory and only a sequence of atomic services, we will use the original action theory or other methods, such as default logic, description logic program and epistemic logic, to reason under an incomplete knowledge of world in the future. At the same time, more control constructors, such as choice and iteration, will be added to enable composing complex services. But, the reducing and reasoning algorithm will also change with those additions.

Acknowledgements. The work is funded by Grant 60873237 from the National Natural Science Foundation of China, and by Grand 4092037 from Beijing Municipal Natural Science Foundation and partially supported by Beijing Key Discipline Program.

References

1. Baader, F.: Description logics. In: Tessaris, S., Franconi, E., Eiter, T., Gutierrez, C., Handschuh, S., Rousset, M.-C., Schmidt, R.A. (eds.) Reasoning Web. LNCS, vol. 5689, pp. 1–39. Springer, Heidelberg (2009)
2. Baader, F., Lutz, C., Milicic, M., Sattler, U., Wolter, F.: A description logic based approach to reasoning about web services. In: The WWW 2005 Workshop on Web Service Semantics (WSS 2005), Chiba City, Japan (2005)
3. Baader, F., Lutz, C., Milicic, M., Sattler, U., Wolter, F.: Integrating description logics and action formalisms: First results. In: 20th National Conference on Artificial Intelligence and the 17th Innovative Applications of Artificial Intelligence Conference, AAAI 2005/IAAI 2005, July 9-13, vol. 2, pp. 572–577. AAAI, Menlo Park (2005)
4. Bhargavan, K., Fournet, C., Gordon, A.D.: A semantics for web services authentication. In: POPL 2004: 31st ACM SIGPLAN-SIGACT Symposium on Principles of Programming Languages, Microsoft Res., Cambridge, UK, January 14-16, vol. 39, pp. 198–209. ACM, USA (2004)

5. Bhargavan, K., Fournet, C., Gordon, A.D., Pucella, R.: TulaFale: A security tool for web services. In: de Boer, F.S., Bonsangue, M.M., Graf, S., de Roever, W.-P. (eds.) FMCO 2003. LNCS, vol. 3188, pp. 197–222. Springer, Heidelberg (2004)
6. Blanchet, B., et al.: An efficient cryptographic protocol verifier based on Prolog rules. In: 14th IEEE Computer Security Foundations Workshop (CSFW-14), vol. 96. Citeseer (2001)
7. Carminati, B., Ferrari, E., Bishop, R., Hung, P.C.K.: Security conscious web service composition. In: 4th IEEE International Conference on Web Services (ICWS), pp. 489–496. IEEE Computer Society, Los Alamitos (2006)
8. Corin, R., Saptawijaya, A.: A logic for constraint-based security protocol analysis. In: 2006 IEEE Symposium on Security and Privacy, Twente Univ., Netherlands, May 21-24, p. 14. IEEE Comput. Soc., Los Alamitos (2006)
9. Kagal, L., Finin, T., Joshi, A.: A policy based approach to security for the semantic web. In: Fensel, D., Sycara, K., Mylopoulos, J. (eds.) ISWC 2003. LNCS, vol. 2870, pp. 402–418. Springer, Heidelberg (2003)
10. Kleiner, E., Roscoe, A.W.: On the relationship between web services security and traditional protocols. In: Mathematical Foundations of Programming Semantics (MFPS XXI) (2005)
11. Levesque, H., Reiter, R., Lesperance, Y., Lin, F., Scherl, R.: Golog: a logic programming language for dynamic domains. Journal of Logic Programming 31(1-3), 59–83 (1997)
12. Reiter, R.: Knowledge in Action. MIT Press, Cambridge (2001)
13. Woo, T.Y.C., Lam, S.S.: A semantic model for authentication protocols. In: IEEE Symposium on Research in Security and Privacy, Dept. of Comput. Sci., Texas Univ., Austin, TX, USA, May 24-26, pp. 178–194. IEEE Comput. Soc. Press, Los Alamitos (1993)

SMAC: Smart Multimedia Archiving for Conferences

Jean Revertera, Maria Sokhn, Elena Mugellini, and Omar Abou Khaled

Abstract. With the advent of new technologies, an increasing amount of (scientific) conferences is being digitally recorded and archived for redistribution. Usually such conferences take the form of a series of talks where different speakers make use of slide-based presentations displayed as a slide-show during the speech. The data within these electronic documents can be used to improve video indexing to facilitating hence the retrieval of specific sequences within a specific video. In order to exploit such data it is however necessary to synchronize the video with the corresponding slide-show presentation. So far such synchronization has been done mainly manually. Nowadays, given the large amount of conferences being recorded, manual archiving is becoming a too time-consuming task that need to be automated. This paper presents an algorithm that automatically segments the video of the presentation and aligns each segment to the corresponding slide. Multiple tests have been done to evaluate the performance of the proposed algorithm and the obtained results, presented in the paper, prove the effectiveness of the proposed algorithm.

Keywords: Video segmentation, Scene detection, Image recognition, Heuristic algorithm, Multimedia synchronization.

1 Introduction

With the expansion of digital video, several research work on video content, video segmentation, image analysis and multimedia synchronization, have been carried out the last decade. While they are mostly designed for general categories of videos, this paper focuses on conference video-recordings.

Jean Revertera · Maria Sokhn · Elena Mugellini · Omar Abou Khaled
University of Applied Sciences of Western Switzerland,
Boulevard Perolles, 80, 1700
e-mail: jean.revertera@hefr.ch, maria.sokhn@hefr.ch
elena.mugellini@hefr.ch, omar.aboukhaled@hefr.ch

E. Mugellini et al. (Eds.): Advances in Intelligent Web Mastering - 3, AISC 86, pp. 143–152.
springerlink.com

The aim in our work is to automatically assign slides (displayed during the talk) to a corresponding video sequence, thus improving their indexation and allowing to provide an improved experience to end-users by making these media (slides and video) interacting together. Several research works have been working in this field: [9, 5, 12, 1, 4]. This novel approach is based on image recognition improved by metadata extracted from the presentation file and on defined heuristic hypothesis.

2 Algorithm Overview

The basic concept is to compare frames extracted from the video with images of the slides extracted from the presentation files. Other prior works have proposed similar systems: [1, 2, 5, 7, 8, 11, 13, 18][1]. This is usually a two step algorithm:

1. The video is divided into segments, each of one represented by one "*key frame*".
2. Each of these segments (ie. frame) is compared against each of the slide pictures (section 5.1).

Our approach distinguish itself by adding a third step, a "matching refinement" stage, where identification results (obtained in point two) are post-processed using metadata extracted from the presentation files and heuristics hypothesis. The resulting workflow is depicted in fig. 1.

The following sections will provide a detailed view of our solution. Section 4 describes the video segmentation, section 5.1 the image identification process, and sections 5.2 to 5.3, the identification refinement algorithm. Section 3 first presents a preliminary consideration about the extracted slide pictures, followed by a detailed description of the algorithm phases.

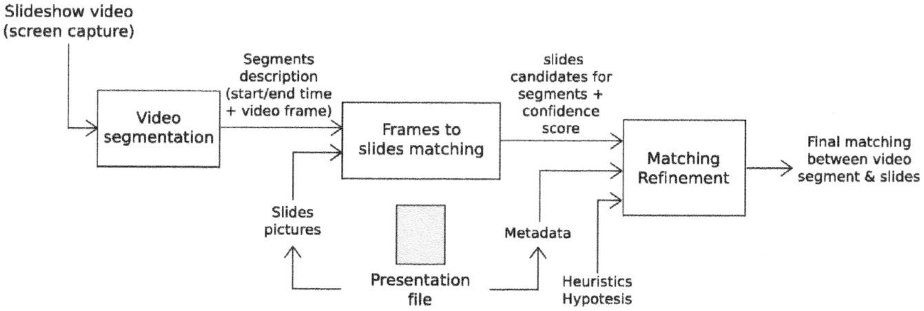

Fig. 1 Overview of the three-phases process

[1] A minority also used speech analysis techniques, like: [4, 9, 12].

3 Preliminary Considerations about Slides Pictures

At any given time, slides pictures may differ from their representation on the screen, mainly because of animation. Slides pictures (as generated by software such as MS-PowerPoint or OpenOffice) are in fact "flattened" view of the slides: all objects are visible AND displayed in their *initial* position (ie. before any moving animation). Since the most used type of animations are sequentially appearing objects (typically text lines), we can postulate the following: the extracted slide pictures will correspond to slides in their *final state* (ie. when all objects are displayed, just before a slide change), except if there is: 1) moving objects (not to be mistaken with objects appearing with a moving transition) 2) object disappearing and not reappearing afterward. These two types of animations could result in what we can call a *unidentifiable final state*. Depending on the case, an approximated match could still be succeeded in the final state, or we may be able to do a match with an *intermediate state* of the displayed slide (when not all animations have been triggered yet) or, in the worst cases, the displayed slide could never be identified. This concept of slides displayed in their intermediate or final state is important and will be used later.

4 Video Segmentation

One of the specificity of our system is that we are using a scan converter which takes the VGA signal from the speaker's computer and multiplexes it into the S-Video input of a capture card and, on the other hand the video projector. In addition, a camera is used to record the speaker's presentation. While our solution requires some additional hardware, it presents the advantage of having a high and constant quality of the slide-show video frames, since all the issues related to the screen area extraction can be ignored (luminosity, geometric deformation caused by the camera position, occlusion of the slide display by the speaker, etc).

The first phase of the algorithm is to detect the slide transitions among the video frames in order to segment the video. Since the video of a slideshow displays mostly static patterns (ie. with a lot of redundant frames) the synchronization is performed on a per-segment basis rather than trying to match each frame individually. The perfect result being that there is exactly one segment for every occurrence of a given slide.

Such transition detection system can be considered as a specific video shot boundary detector. Video shot boundary detection is a mature field in video analysis and includes several available algorithms, as the ones stated in [3, 10, 18]. However, the cited algorithms are designed for generic video sequences and not for slideshows, which are mainly composed of static frames, with quick hard cuts, involving only a small number of pixels change (all slides in a presentation generally present a very homogeneous style). Thus the necessity of developing a specialized system (this thought is shared by [17]).

In such systems, false-positive (spurious detection) will often occurs due to animations in the displayed slide causing activity on the screen. On the other hand, if the sensibility of the system is not high enough, false-negative (slide changes not detected), may also occurs. False-positives result in redundant or non-identifiable video frame, which could usually be filtered in a second time (ie. when proceeding with the identification of frames). Conversely, false-negatives are nearly impossible to fix, since they are very difficult to detect. Therefore the algorithm has to be sensitive enough to remove a maximum of false-negative, even if it does imply more false-positives (this conclusion is shared by [13]).

We proceed as follows: first the absolute difference between two adjacent frames is computed and the resulting values are stored into three 2D-matrixes, one for each color channel. Then we calculate the variance[2] of each matrix, these three values are then summed and compared to a pre-defined threshold: if it's greater, a transition is detected. This threshold characterizes the sensitivity of our system.

In order to avoid an overload of transitions detection (eg. when a video is running on the screen or during an off-slide demonstration), we stop storing video frames when an excessive number of transitions has been detected in a slot of time and we start capturing frame again as soon as a stable state is retrieved.

Finally, for each found segment, a significant frame has to be chosen in order to be compared later with the slide pictures. It has been decided to simply take the last one, since it is the more likely representing a slide in its final state.

5 Matching Refinement

5.1 Frames Identification

Reviewed papers mostly presented their own solution to the issue of image comparison. Some of them used DCT [5], other SIFT [8], while some other tried to use combinations of various features [2]. All of them presented fairly good results. We choose an algorithm which has already been proved successful in a case of use close to ours: an edge histogram detection specialized for text content. Should you need more detail, the paper [7] presents the detailed process.

5.2 Identification Refinement

A large part the slide show video ($> 60\%$ in some cases) could represents frames which do not correspond to the pictures we have extracted from the presentation file (cf. table 1, p.150), this is due to animations or external demonstration. This gives places to a large amount of segments where only few of them will have an usable identification.

Once the video segmentation is available as well as the features vectors from the significant video frames, and the ones from the slide pictures extracted from the presentation file (using the algorithm presented in section 5.1).

[2] The variance provide a more convenient value for the thresholding, and allow us to ignore uniform alteration between two frames which may be due to noises.

1. Comparisons of features vectors: The system compares each frame features vector with all the slides features vectors. The results are stored into an internal data structure.
2. Deduction of "candidates": all segments have at least one slide candidate, the one producing the best (lowest) difference score when compared with a segment frame. When the result of the identification is ambiguous, additional candidates could be kept by inducing a concept of "relative confidence" of the identification. In order to be kept, the i-th additional candidate (starting from the second best matching slide to the worst one) have to have a relative confidence C_i greater than a given threshold. With C_i being:

$$C_i = \frac{s_i - s_0}{s_0} \quad (1)$$

   With S_i: the difference score of the i-th candidate, and S_0 the difference score of the best candidate. Segments providing a high confidence identification will likely have a single candidate for identification (the best one).
3. Filtering non-usable frames: Most probably, some selected frames do not match with any slides (eg. during external demonstration), they will likely present a low-confidence with many available candidates (ie. the difference scores are uniformly distributed, without strongly advantaging any slide).
4. Deduction of the base sequence: The key heuristic employed here, is to consider that slides are sequentially displayed from the first slide to the last one: :
   - The chosen segments are ordered sequentially in time.
   - Identified slides numbers have to be sequential with respect to the segment order.
   - The sum of the concerned identifications confidence has to be maximized.

   This concept of base-sequence provide two interesting features: 1) if an identification was unsuccessful, we still have a chance to get the right candidate kept (ie. not forcibly the candidate which had the lowest difference score) 2) the members of the base sequence are identified with a higher reliability: so a future occurrence of a frame identification result should have a features vector quite similar.
5. Add non-sequential displayed slides: Non-sequential slides occur when the presenter goes backward in the slide stream, (and are therefore not contained in the base sequence). A non-sequential slide is added to the intermediary result only if the matching has a good confidence and if the sequence last more than five seconds. Adjacent sequences assigned to a same slide are merged.
6. Add remnant slides: Remnant slides are non-sequential slides which do not fulfill the condition listed in the previous step. In practice, they only need a lower confidence and don't have to last a minimal amount of time to be kept, but they matched the same slide id as an adjacent already assigned sequence. This could happen when a slide is in an intermediary state and provides a non-perfect identification (ie. with a low confidence). Adjacent segment with redundant slide assigned are merged.
7. Assign orphan sequences: Addressed in the next section (5.3).

5.3 Orphan Sequences Assignment

There is some case where a video segment could not be matched to a slide, this may be due to the animations displayed, or because of an external demonstration (eg. opening a web browser) occurred. These non-identifiable sequences which remain after the first phase of matching refinement are called "orphan sequences".

As described in section 3, a segment matching is likely to be succeeded when a slide is displayed in its final state. This allows us to define a default behavior when in presence of orphan sequences: most of time they have to be assigned to the same slide than their *following*, formally identified, neighbor. This default behavior already works well in the majority of the cases, but we can further enhance it by using some meta-data related to the presentation.

To do this, we extract a set of internal information from the presentation file[3], more particularly: we're interested in knowing which slides show animations, and which are statics. In practice, it is more reliable to define which slides are statics (they simply don't have any internal object which may be animated), than to define which slides are animated (for example, PNG/GIF pictures could be animated or not, an animation could be located outside the canvas of the slide, and so on). Starting from this, a set of four rules has been defined, regrouping the four possible case of figures (legend: PS = previous sequence, FS = following sequence):

- The previous and the following identified sequences are static slides. The orphan sequence is very likely an off-slide demonstration, and most probably linked to the previous sequence (a demonstration is almost always related to the current slide, which may sometimes be just an introduction to the demo itself): *Assignment to PS.*
- The previous sequence is static and the following is not. The orphan sequence could be present due to an off-slide demonstration, or an animation related to the following sequence. External demonstrations being naturally less frequent than animations: *Assignment to FS.*
- The following sequence is static and the previous is not. The orphan sequence presence is due to an animation in the previous sequence or an external demonstration. In both case of figure: *Assignment to PS.*
- The previous and the following identified sequences are non-static slides. The orphan sequence presence can be due to animations either in previous or following sequences, or to an off-slide demonstration. We fall back to default most likely behavior (identifications succeeded with slides in their final states): *Assignment to FS.*

6 Results

The platform which serves as a base for this system is SMAC[4], a conference capturing system similar to [14] and [16]. An example of interface is showed on fig. 2.

[3] We used XSL stylesheets on the internal XML files from MS-Powerpoint 2007 and Open-Document presentation files.

[4] Smart Multimedia Archive for Conferences: `http://smac.hefr.ch/`

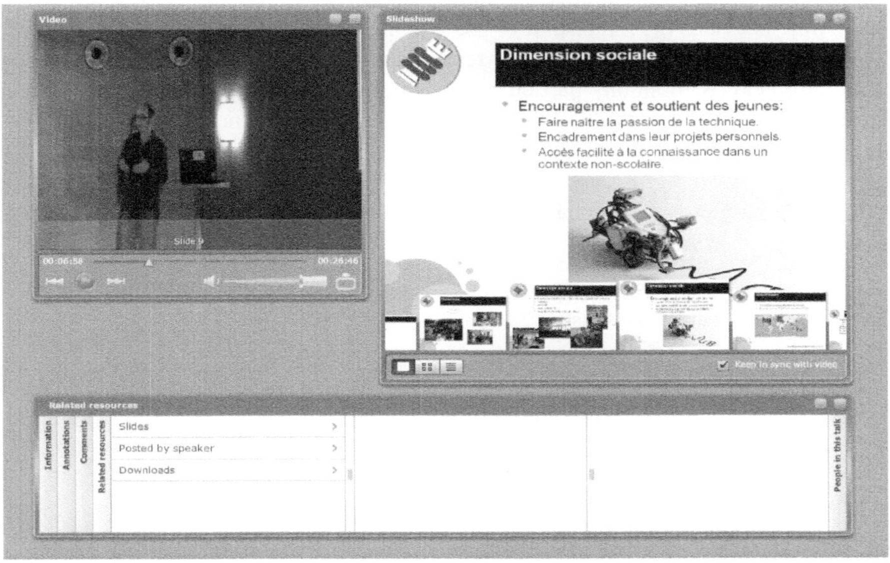

Fig. 2 Video recording visualisation interface

Our testbed is composed of sixty conferences, mainly provided by CERN[5]. Each conference lasts twenty up to sixty minutes. All time values were observed on a bi-processor Intel Pentium D 2Ghz, using a Windows-XP platform with 2Go of RAM. The implementation language is Python with some low-level APIs. All images (frames and slides) are processed in the lossless PNG format.

6.1 Video Transitions Detection Result

The video segmentation system was tested with a detection rate of one frame per second of video. The system performed exactly as planned: no false-negative on slide changes, and the filtering of the most of the non-usable false-positive (cf. value Q_f in the result table).

The measured time of execution, for a presentation of one hour long, is about sixteen minutes for 3600 frame (the presentation real time * 0.27).

6.2 Matching Refinement Result

Table 1 presents the final results of our algorithm on ten representative presentations. On the sixty presentations, only two of them (contributions 9 and 10 as showed in the result table) contain a falsely synchronized segment. It was each time caused

[5] European organization for nuclear research.

by some imprecision at the level of the orphan sequence assignment (which in the same time provides the major part of the synchronization).

One thing that these results on our testbed show clearly, is that an identification-only approach (without any particular refinement in a second time) is clearly non-sufficient for video having less than 50% of their frames which graphically corresponds to a given extracted slide pictures, proving the importance of the applied refinement.

The Matching refinement takes in general no more than one second to be applied (an average of 929ms).

Table 1 Global results and statistics

Contrib.	T_i	T_{nia}	T_{nio}	S_n	S_c	Q_f	Q_n (Q_s/Q_a/Q_e)	G_{id}	M_{no}	Err_{no}	M_o	Err_o
1	$5'03''$	$3'55''$	$0'$	12	14	0	41 (14/27/0)	25	23	0	18	0
2	$18'05''$	$3'58''$	$0'$	15	17	0	27 (17/10/0)	21	22	0	5	0
3	$19'27''$	$0'$	$0'$	19	18	0	18 (18/0/0)	18	18	0	0	0
4	$29'08''$	$0'$	$0'$	23	25	0	25 (25/0/0)	25	25	0	0	0
5	$34'51''$	$0'$	$0'$	20	20	0	20 (20/0/0)	20	20	0	0	0
6	$15'45''$	$1'34''$	$0'$	15	24	0	28 (24/4/0)	28	28	0	0	0
7	$18'50''$	$3'58''$	$0'$	21	23	0	43 (23/20/0)	30	35	0	8	0
8	$22'26''$	$0'30''$	$0'$	24	24	0	26 (24/2/0)	24	24	0	2	0
9	$11'11''$	$3'34''$	$3'23''$	15	19	255	103 (19/52/32)	19	35	0	68	6
10	$8'12''$	$16'22''$	$0'$	29	29	135	112 (29/83/0)	56	52	0	60	4

Description:

- T_i: time during which the video displays a slide in an *identifiable state*, ie. when the displayed content corresponds to a given slide picture.
- T_{nia}: time during which a displayed slide on the video doesn't match the picture extracted from the presentation file.
- T_{nio}: time during which the video displays a non-slide content (eg. web browser).
- S_n: the number of slides in the presentation.
- S_c: the number of slide changes which effectively occurred. It could be $> S_n$ (at least one slide was displayed several time because of a backward iteration in the slide stream) or $< S_n$ (at least one slide wasn't displayed at all).
- Q_f: the number of sequences filtered by the passive mode.
- Q_n: the total number of detected sequence. It's composed from Q_s: the number of sequences related to slide changes, Q_a: the number of sequences related to in-slide activities (animations) and Q_e: the number of sequences related to off-slide activities.
- G_{id}: the number of sequences correctly identified.
- M_{no}: the number of sequences matched to a specific slide *without* orphan sequences assignment.
- Err_{no}: the number of falsely matched sequences in M_{no}.
- M_o: the number of sequences matched through orphan sequences assignment.
- Err_o: the number of falsely matched sequences in M_o.

7 Conclusion and Future Works

We presented in this paper a novel approach to align a slide show video with the corresponding slides pictures extracted from the presentation file, by successfully combining video-segmentation, pattern recognition, metadata handling, and heuristics. Although not being entirely failsafe (we have seen that actually only a small percentage video frames will graphically corresponds to a slide picture in a representative presentation), this system shows quite satisfying results, and could be used in a way than manual reviewing is only occasional.

There is however still room for improvement. In our experiment we only used the presentation file metadata to check whenever there is or not animations on each slides, but we could potentially extract much more useful information (eg. whenever the animation is appearing or disappearing, the area of the screen affected, etc.), this could led to further improve orphan sequence alignement. We could also consider exploiting the audio track of the video, using voice recognition or (more likely in the near future) silence detection in order to provide a more fine-grained temporal alignment (eg. the timing of a slide change could be adapted in order to match a break in the speaker speech). The prototype will be further tested on a weekly basis.

References

1. Syeda-Mahmood, T.F.: Indexing for topics in videos using foils. In: IEEE Conference on: Computer Vision and Pattern Recognition, vol. 2, pp. 312–319 (2000)
2. Behera, A., Lalanne, D., Ingold, R.: Looking at projected documents: event detection and document identification. In: IEEE International Conference on Multimedia and Expo., vol. 3, pp. 2127–2130 (June 2004)
3. Boreczky, J.S., Wilcox, L.D.: A hidden markov model framework for video segmentation using audio and image features. In: IEEE International Conference on Acoustics, Speech and Signal Processing, vol. 6, pp. 3741–3744 (1998)
4. Chen, Y., Heng, W.J.: Automatic synchronization of speech transcript and slides in presentation. In: Proceedings of the 2003 International Symposium on Circuits and Systems, vol. 2, pp. 578–651 (2003)
5. Chiu, P., Foote, J., Girgensohn, A., Boreczky, J.: Automatically linking multimedia meeting documents by image matching. In: Proceedings of the Eleventh ACM on Hypertext and Hypermedia, pp. 244–245 (2000)
6. Daddaoua, N., Odobez, J., Viniciarelli, A.: Ocr based slide retrieval. In: Proceedings of the Eighth International Conference on Document Analysis and Recognition, pp. 945–949 (2005)
7. Erol, B., Hull, J.J., Lee, D.S.: Linking multimedia presentations with their symbolic source docuements: algorithm and applications. In: Proceedings of the Eleventh ACM International Conference on Multimedia, pp. 498–507 (2003)
8. Fan, Q., Barnard, K., Amir, A., Efrat, A., Lin, M.: Matching slides to presentation videos using sift scene background matching. In: POSTER SESSION: Poster Session 2: Annotation, Summarization, and Visualization, pp. 239–248 (2006)
9. Franklin, D., Bradshaw, S., Hammond, K.J.: Jabberwocky: you don't have to be a rocket scientist to change slides for a hydrogen combustion lecture. In: Intelligent User Interfaces, pp. 98–105 (2000)

10. Lienhart, R.: Comparison of automatic shot boundary detection algorithms. In: Storage and Retrieval for Still Image and Video Databases IV, pp. 170–179 (1996)
11. Liu, T., Hjelsvold, R., Kender, J.R.: Analysis and enhancement of videos of electronic slide presentations. In: Proceedings of 2002 IEEE International Conference on Multimedia and Expo., vol. 1, pp. 77–80 (2002)
12. Martin, T., Boucher, A., Ogier, J.M., Rossignol, M., Castelli, E.: Multimedia scenario extraction and content indexing for e-learning. In: International Workshop on Content Based Multimedia, pp. 204–211 (2007)
13. Mukhopadhyay, S., Smith, B.: Passive capture and structuring of lectures. In: Proceedings of the Seventh ACM International Conference on Multimedia (Part 1), pp. 447–487 (1999)
14. U. of Michigan Media Union and CERN: Web lecture archive project, `http://www.wlap.org`
15. I. organisation for standardisation: Mpeg-7 overview (October 2004), `http://www.chiariglione.org/mpeg/standards/mpeg-7/mpeg-7.htm`
16. I. Sonic Foundry: Mediasite, `http://www.sonicfoundry.com/mediasite`
17. Wang, F., Ngo, C.W., Pong, T.C.: Synchronization of lecture videos and electronic slides by video text analysis. In: Proceedings of the Eleventh ACM Internation Conference on Multimedia, pp. 315–318 (2003)
18. Yuan, J., Wang, H., Xiao, L., Zheng, W., Li, J., Lin, F., Zhang, B.: A formal study of shot boundary detection. IEEE Transactions on Circuits and Systems for Video Technology, 168–186 (2007)

Ontological-Based Information Extraction of Construction Tender Documents

Rosmayati Mohemad, Abdul Razak Hamdan,
Zulaiha Ali Othman, and Noor Maizura Mohamad Noor

Abstract. Extracting potentially relevant information either from unstructured, semi structured or structured information on construction tender documents is paramount with respect to improve decision-making processes in tender evaluation. However, various forms of information on tender documents make the information extraction process non trivial. Manually identification, aggregation and synthesize of information by decision makers is inefficient and time consuming. Thus, semantic analysis of content and document structure using domain knowledge representation is proposed to overcome the problem. The ontological-based information extraction processes contain three important components; document structure ontology, document preprocessing and information acquisition. The findings are significantly good in precision and recall which the performance measures have reached accuracy of precision about 82.35 % (concepts), 96.10 % (attributes), 100% (values) and 100 % of recall for both parameters of concepts and attributes, while 91.08 % for values.

Keywords: Document Analysis, Information Extraction, Ontology, Construction Tender.

1 Introduction

Nowadays, ontology plays an essential role in knowledge management in which it shares common understanding of a domain. Since the last decade ontology has

---

Rosmayati Mohemad · Abdul Razak Hamdan · Zulaiha Ali Othman
Faculty of Information Science and Technology,
Universiti Kebangsaan Malaysia, 43600 Bangi,
Selangor Darul Ehsan, Malaysia
e-mail: {arh,zao}@ftsm.ukm.my

Rosmayati Mohemad · Noor Maizura Mohamad Noor
Department of Computer Science, Faculty Science and Technology,
Universiti Malaysia Terengganu, 21030 Kuala Terengganu,
Terengganu Darul Iman, Malaysia
e-mail: {rosmayati,maizura}@umt.edu.my

E. Mugellini et al. (Eds.): Advances in Intelligent Web Mastering - 3, AISC 86, pp. 153–162.
springerlink.com

achieved a great success in multiplicity of research fields including e-commerce [1,2], biomedical [3], bioinformatics [4] and others. However, knowledge management in construction tender has been less intensively studied due to its additional challenges in information integration, information analysis and weak interoperability between stakeholders.

Construction tendering processes involve large volumes of complex text documents in paper-based and digital format [5]. Tender documents may consist of unstructured (terms of contract in natural language sentences), semi structured (form-based) and structured (tabular) information. These documents involve diversity of information such as project specifications, terms and regulations of contract, tendering procedures, tender forms and supporting documents [6]. In addition, tenderers usually tend to provide manifold of documents to prove their abilities to win the tender. All these complicated features contribute as the challenges to automated information extraction since the documents are purposely designed for human-readable. Therefore, extracting information from text documents requires knowledge of both document structure and contents.

In construction tender, recognizing relevant information from tender documents is necessary for decision-making process, especially in tender evaluation based on multi-criteria [6]. Current approach is impractical and time consuming when the evaluators need to identify, aggregate and synthesize salient information of these criteria manually. Subsequently, information that would be useful for making a decision may be missing. This paper proposed an approach to computationally extract and map human-readable document structure information into machine readable format using ontology. Here, predefined key knowledge automates the task of extracting tender information and hence improving information finding. The approach can be customized as general information extraction tool for other similar document structure.

Ontology is a knowledge representation model of real concepts and intricate relationships between those concepts. We explore semantic-based representation between concepts and relationships between them based on common vocabulary of document structure. Further, complex concepts are also built from simpler concept definitions using OWL operators including union, intersection and complement. The extracted information is matched and associated with domain specific keywords and regular expression data patterns defined in the ontology. Users are permitted to query extracted data and infer logical rules using Pellet OWL to recognize which concepts fit under which definitions. Based on rules inferred, the possible information is located to be extracted when it satisfies the rule defined. In this way, the meaning of a document is recognized and significant information will be available for evaluation process.

The paper proceeds in the following manner. Firstly, related researches on ontological-based information extraction are briefly reviewed in related works of Section 2. Section 3 describes about proposed ontological-based information extraction processes. Next, the experimental setup and results are provided and discussed in Section 4 and Section 5 respectively. Finally, Section 6 concludes with summary of this paper and future research directions.

2 Related Works

The importance of ontology is recognized and implemented in information extraction. Information extraction is defined as any method of analyzing large volumes of unstructured texts, normally in the form of natural language and automatically extracts salient information from these texts into pre-defined template [7-9]. Ontology and information extraction are closely related in two main tasks either pre-existing ontology is used to extract information or it is used to populate and improving ontology. The first task is our focus in the research. The use of ontology enables information extraction to have better access and coverage to relevant information by providing domain knowledge specific application [10,11]. Output from this process is useful for further processing in wide range of application such as text mining, text classification, ontology learning, information retrieval, decision support systems and others.

Research that was done by Holzinger et al. [12] populated domain specific ontology with instances extracted from structural information on tabular data of Web documents. Here, they come out with table ontology and fixed adjacent attribute-value pairs to identify the instances. The goal of our paper shares the same interest of instantiating domain ontology, yet expands to process non structural and semi structural information. Table ontology is improved by identifying headers of tabular structure and overcome fixed adjacent attribute-value pairs approach. In addition, non tabular concepts also included to give semantic knowledge for non tabular data. Meanwhile, Shashirekha and Murali [13] improved similar framework that had been proposed by Embly et al. [14] by identifying relevant information written in short forms or abbreviations using domain dependent ontology and populated the extracted information into relational database. Nevertheless, they examined only non structural documents and the semantic of document structures were not considered. In our approach, the semantic representations allow information to be extracted directly from documents and can be queried directly from the ontology.

WeDax is web-based data extraction tool where it restructures web documents into XML schema representations and mapping with domain specific ontology [15]. The tool however only managed to extract constantly changing data with a fixed structure. Moreover, XML is more to syntactic language, thus lack supports for efficient sharing between semantically defined concepts. Instead, our research uses the classification capabilities of rule-based inference engine to identify meaningful instances without having to depend on conventional mapping process. Other information extraction application has been proposed by Biletskiy et al. [16] is Course Outline Data Extractor. The tool used data integration to automatically transform learning syllabus stored on HTML web pages into XML format. Then, the relevant information is extracted by computing similarity between source and target syllabus in XML using domain specific ontology. However, the main different of these researches is they do not consider enough semantic relations among fundamental concepts of document structure and recognize instances depend on heuristic mapping and matching algorithms.

3 Ontological-Based Information Extraction Processes

In this paper, ontology is used to give semantic analysis of document concepts and their instances in order to assist information extraction on tender documents. Hence, the processes of ontological-based information extraction as shown in Fig. 1 are proposed. The processes generally involve preprocessing of PDF tender documents, then possible relevant information is identified through mapping and matching before the recognized information is stored in the document structure ontology. Next subsections discuss in detail on document structure ontology, document preprocessing and information acquisition.

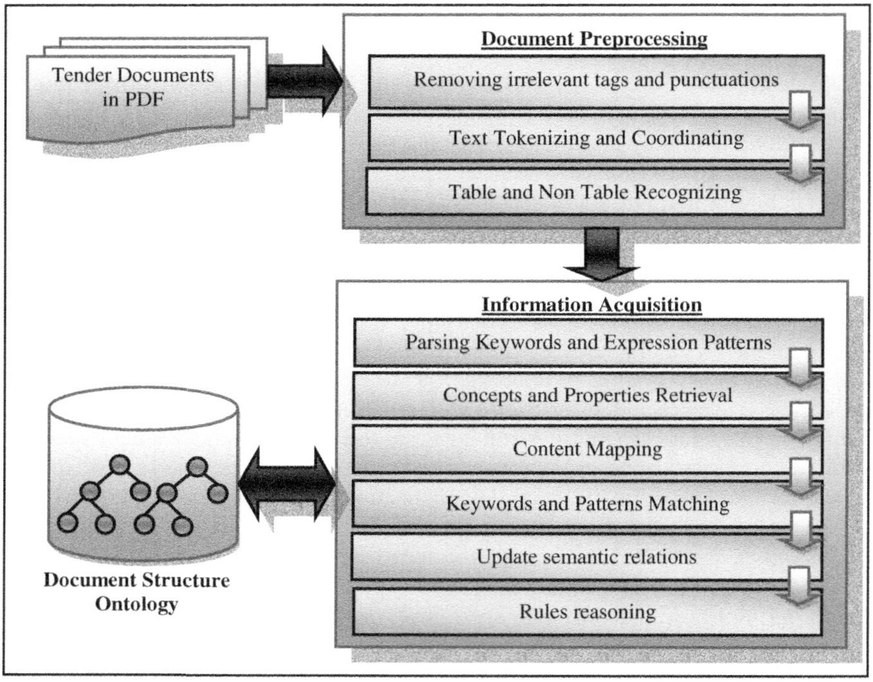

Fig. 1 Ontological-based Information Extraction Processes

3.1 Document Structure Ontology

Ontological modeling of document represents knowledge about the structure of construction tender document considering most information on the document are visually represented as full sentences, form-based and tabular data. The purpose of document structure ontology is to provide semantic knowledge representation on each concept in a document. Fig. 2 shows ontological modeling of basic concepts such as *Document*, *Non-Table*, *Table*, *Paragraph*, *Column*, *Row*, *Cell*, *Header*,

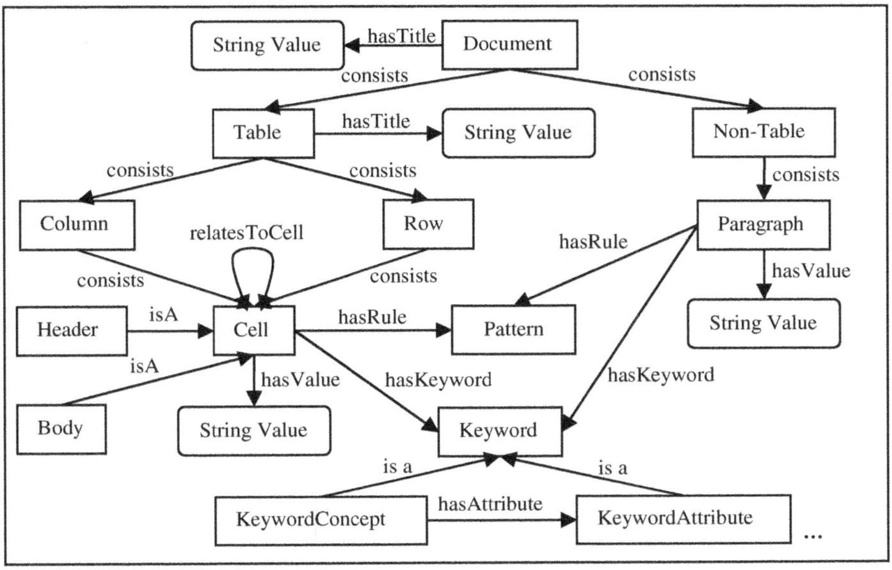

Fig. 2 Document Structure Ontology

Body, *Keyword*, *Pattern* and these concepts are associated with relationships. Some concepts in the ontology are reused from table ontology as proposed by Holzinger et al. [12]. The ontology populates instances according to the structure found on the document. Each document and table is reflected by *Document* and *Table* concepts respectively in which are differentiated by string title. Transitive *consists* relation is modeled as OWL object property which indicates semantic relationship between concepts of *Document*, *Table*, *Row*, *Column*, *Cell*, *Non-Table* and *Paragraph*. Relevant data or value found is stored into either *Cell* or *Paragraph* which contains string value modeled as OWL data type property. This value is associated with matched keywords and pattern rules. Concept of *Keyword* can be inherited by subclasses such as *KeywordConcept*, *KeywordAttribute* and etc. Particular concepts and attributes are represented using predefined keywords and are associated with *hasAttribute* relationship. Here, related keywords and regular expression patterns are defined to guide the extraction. Meanwhile, complex concepts can be derived from basic concepts definitions. All of these concepts and relationships are encoded using the most recent standard ontology language of OWL. The document model can represent any standard document that contains three different type of information either non structured, semi structured or unstructured. It is adequate in modeling construction tender document structure.

3.2 Document Preprocessing

Initially, original documents are preprocessed using special tasks such as removing irrelevant tags and punctuations, text tokenizing and coordinating. Tokenizing splits all sentences into single word. Identification of coordinate (x, y) for each token on the documents is essential since all the documents are in non structural Portable Document Format (PDF). Table structure in PDF documents do not have any identified tagged characteristics in common. This operation is accomplished using JPedal architecture (http://www.jpedal.org). Furthermore, table extraction algorithm is applied to recognize tabular structure and tabular content. Also, the algorithm is able to identify non tabular text. The preprocessed tabular and non tabular text data are transformed into specific text representation vector matrix. Both types of text represent information in unstructured, semi structured and structured form.

3.3 Information Acquisition

Structural relationships of documents are expressed by applying document structure ontology. Initially, basic concepts and properties defined in the document structure ontology is parsed. This also includes all the predefined keywords and regular expression data patterns. Subsequently, content mapping algorithm maps related document structures and data values found in preprocessed text as instances into ontology. Concept definitions on the ontology specify document structures and instances as data values. Semantic relationships between these instances are recognized as well. Possible semantic relationships between identified instances are updated by matching with lexical value of keywords and regular expression patterns. As the result, the ontology derives all facts of document structures and data values. In order to interpret and analyze the meaning of extracted content, OWL Pellet inference engine is used to reasoning rules derived in concepts. In this way, simpler concept definitions infer complex concept definitions

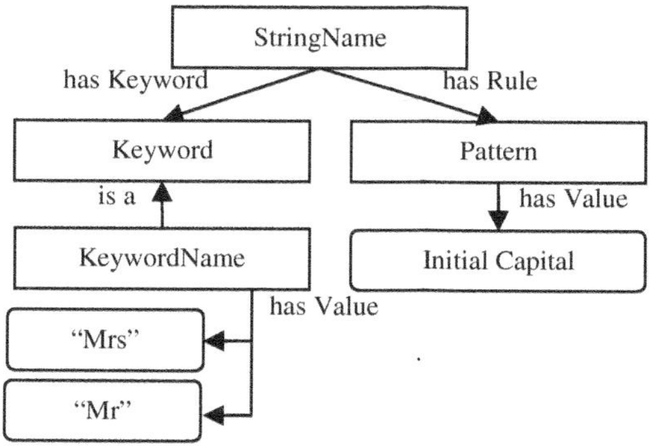

$StringName = \forall Cell \cap \exists.hasKeywordKeywordName \cap \exists.hasRulePattern$

Fig. 3 Example of Complex Class Using OWL Operator *and*

using OWL operators. For example, one of the possible rules that can be inferred by OWL Pellet reasoner is depicted in Fig. 3. Here, complex class of *StringName* is derived by union of any *Cell* has *KeywordName* and has *Pattern*. Furthermore, SPARQL query is executed to retrieve the extracted information. Concepts of domain specific keywords identification and regular expression patterns are semantically associated with cell and paragraph.

4 Experimental Setup

The purpose of this experiment is to evaluate the performance of the proposed approach according to precision and recall measurements. Six copies of tender

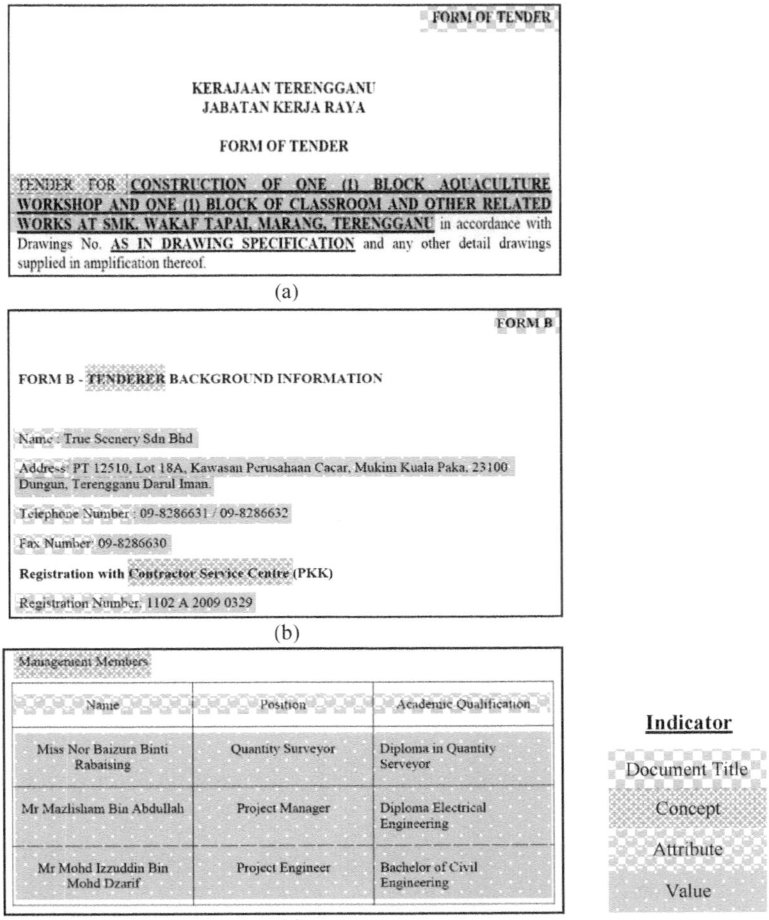

Fig. 4 Types of Information on Construction Tender Documents; a) Unstructured Information, b) Semi Structured Information, and c) Structured Information

documents of similar building construction project based on Malaysia Construction Tender are used as the experimental data. The average pages per document is approximately fifteen pages. Each document consists of compulsory information about tender agreement, certified approval letter, contractor background, financial data, technical staff, list of construction plant and equipment, past and current project of six different contractors. These information are visually represented in natural language sentences, form-based and table.

Fig. 4 presents the sample of tender documents format with three different types of information. Each document generally contains document title, concepts, attributes and values. The information in natural language sentence is considered as unstructured that depicted in Fig. 4a. The highlighted areas are concept and adjacent value of the particular concept that need to be extracted. Meanwhile, Fig. 4b shows the semi-structured information where the relevant information that need to be identified is concepts, attributes that relate to the concepts and adjacent values. Fig. 4c represents an example of structured information where the information is organized in tabular format. The experiments are run in Java-based environment and divided into three strategies according to information types.

5 Result and Evaluation

The experiment was run to extract details content about tender such as tender title, bid price and time to complete a project included as unstructured information, form-based representation of tenderer background profile and structured table denotes data on company staff, list of facilities available, financial record, current projects and past projects.

In order to evaluate the extraction accuracy result, standard information extraction method of precision, recall and f-measure have been applied. Table 1 shows the comparison results of both evaluation methods for computerized extraction. Three parameters that have been evaluated are concepts, attributes and values. These parameters reflects the prime categories of information that need to be extracted. The evaluation of precision, recall and f-measure have shown significantly good accuracy in detecting relevant information. The precision rates for concepts, attributes and values have reached 82.35 %, 96.10 % and 100 % respectively. The recall have achieved 100 % for both parameters of concepts and attributes, while 91.08 % for values. Meanwhile, tests accuracy of f-measure are 90.32 % for concepts, 98.01 % for attributes and 95.33 % for values.

Table 1 Comparison Results of Information Extraction based on Precision, Recall and F-Measure

	Computerized Information Extraction		
	Precision	**Recall**	**F-Measure**
Concepts	82.35 %	100 %	90.32 %
Attributes	96.10 %	100 %	98.01 %
Values	100 %	91.08 &	95.33 %

The finding shows that the ontology is significantly capable in recognizing important of concepts, attributes and values and then represent them as instances into machine readable format of ontology that can be used for further decision-making process. The results indicate the approach is capable to detect information that matched with the predetermined keywords and regular patterns. In addition, the capability of ontology to allow rules reasoning improve the extraction process.

6 Conclusion

This study has proposed an approach of information extraction using domain dependent ontology. The relevant information is recognized by matching with keywords and regular expression patterns. Based on precision, recall and f-measure, the extracted information has shown significant experimental results. However, the implementation of keywords and regular expression patterns allows any matched strings to be associated with them including two different strings that contain similar word. There is a chance to produce redundant recognizable information and ambiguous interpretation of the knowledge. In future, the meanings of each keyword will be hierarchical expanded in the ontology in order to produce more quality keywords. Furthermore, table recognizing which currently works based on simple table assumption will be enhanced for complex table. We are considering to include supporting documents as part of document sources and proposed a model to semantically match between the content of compulsory and supporting documents.

References

1. Kayed, A., Colomb, R.M.: Extracting Ontological Concepts for Tendering Conceptual Structures. Data & Knowledge Engineering 40(1), 71–89 (2002a)
2. Du, T.C.: Building an Automatic e-Tendering System on the Semantic Web. Decision Support Systems 14(1), 13–21 (2009), doi:10.1016/j.dss.2008.12.009
3. McCray, A.T.: An Upper-Level Ontology for the Biomedical Domain. Comparative and Functional Genomics 4(1), 80–84 (2003)
4. Schulze-Kremer, S.: Ontologies for Molecular Biology and Bioinformatics. Silico Biol. 2(3), 179–193 (2002)
5. Soibelman, L., Wu, J., Caldas, C., Brilakis, I., Lin, K.-Y.: Management and Analysis of Unstructured Construction Data Types. Advanced Engineering Informatics 22(1), 15–27 (2008)
6. Rosmayati, M., Abdul Razak, H., Zulaiha, A.O., Noor Maizura, M.N.: Ontological-based for Supporting Multi Criteria Decision-Making. In: Wen, D., Zhou, J. (eds.) 2nd IEEE International Conference on Information Management and Engineering, Chengdu, China, pp. 214–217. IEEE Press, Los Alamitos (2010)
7. Cowie, J., Lehnert, W.: Information Extraction. Communications of the ACM 39(1), 80–91 (1996)
8. Soderland, S.: Learning Information Extraction Rules for Semi-Structured and Free Text. Machine Learning 34(1-3), 233–272 (1999)

9. Grishman, R.: Information Extraction: Techniques and Challenges. In: Pazienza, M.T. (ed.) SCIE 1997. LNCS, vol. 1299, pp. 10–27. Springer, Heidelberg (1997)
10. Maedche, A., Neumann, G., Staab, S.: Bootstrapping an Ontology-Based Information Extraction System. In: Szczepaniak, P., Segovia, J., Kacprzyk, J., Zadeh, L. (eds.) Intelligent Exploration of the Web. Studies In Fuzziness And Soft Computing, pp. 345–359. Springer/Physica-Verlag, Heidelberg (2003)
11. Alani, H., Kim, S., Millard, D.E., Weal, M.J., Hall, W., Lewis, P.H., Shadbolt, N.R.: Automatic Ontology-Based Knowledge Extraction from Web Documents. IEEE Intelligent Systems 18(1), 14–21 (2003)
12. Holzinger, W., Krüpl, B., Herzog, M.: Using Ontologies for Extracting Product Features from Web Pages. In: Cruz, I., Decker, S., Allemang, D., Preist, C., Schwabe, D., Mika, P., Uschold, M., Aroyo, L.M. (eds.) ISWC 2006. LNCS, vol. 4273, pp. 286–299. Springer, Heidelberg (2006)
13. Shashirekha, H.L., Murali, S.: Ontology Based Structured Representation for Domain Specific Unstructured Documents. In: Proceedings of International Conference on Conference on Computational Intelligence and Multimedia Applications 2007, Tamil Nadu, pp. 50–54 (2007)
14. Embley, D.W., Campbell, D.M., Smith, R.D.: Ontology-Based Extraction and Structuring of Information from Data-Rich Unstructured Documents. In: Proceedings of the Conference on Information and Knowledge Management (CIKM 1998), Washington D.C, pp. 52–59 (1998)
15. Snoussi, H., Magnin, L., Nie, J.-Y.: Toward an Ontology-based Web Data Extraction. In: Proceedings of the 15th Canadian Conference on Artificial Intelligence, Calgary, Altas, Canada, pp. 1–8 (2002)
16. Biletskiy, Y., Brown, J.A., Ranganathan, G.: Information Extraction from Syllabi for Academic e-Advising. Expert Systems with Applications 36(3), 4508–4516 (2009)

Using Level-2 Fuzzy Sets to Combine Uncertainty and Imprecision in Fuzzy Regions

Verstraete Jörg

Abstract. In many applications, spatial data need to be considered but are prone to uncertainty or imprecision. A fuzzy region - a fuzzy set over a two dimensional domain - allows the representation of such imperfect spatial data. In the original model, points of the fuzzy region where treated independently, making it impossible to model regions where groups of points should be considered as one basic element or subregion. A first extension overcame this, but required points within a group to have the same membership grade. In this contribution, we will extend this further, allowing a fuzzy region to contain subregions in which not all points have the same membership grades. The concept can be used as an underlying model in spatial applications, e.g. websites showing maps and requiring representation of imprecise features or websites with routing functions needing to handle concepts as *walking distance* or *closeby*.

1 Introduction

The concept of the fuzzy regions originated from a need to represent and reason with imperfect spatial information. The available models did not provide ample capabilities deal with imprecision or uncertainty of the data, particularly when the imprecision or uncertainty concerned the locational data itself: a region without a crisply defined outline, a region with elements that only partly belong to it or a point located at an imprecise location. The concept of fuzzy regions solved this, the fuzzy set over the two dimensional domain can be given a veristic interpretation ([1]),

Verstraete Jörg
Systems Research Institute - Polish Academy of Sciences,
Ul. Newelska 6, 01-447 Warszawa, Poland
Database, Document and Content Management,
Department of Telecommunications and Information Processing,
Sint Pietersnieuwstraat 41, 9000 Gent, Belgium
e-mail: jorg.verstraete@ibspan.waw.pl,
jorg.verstraete@telin.ugent.be

E. Mugellini et al. (Eds.): Advances in Intelligent Web Mastering - 3, AISC 86, pp. 163–172.
springerlink.com

where the membership grades indicate the extent to which points belong to the set thus representing a region; or a possibilistic interpretation ([1]) where the membership grades indicate the possibility this point is a valid candidate, thus indicating a fuzzy point. However, in the model, all elements were considered independently from one another, yet sometimes a user can have additional information. An example would be the representation of a lake with a changing water level. All points at the same altitude along the side of the lake will be either above the water or under the water at the same time; so it makes sense to group these points together if we want to represent the lake as a fuzzy region.

In this contribution, we will go deeper into a mechanism that allows such internal dependencies to be modelled. In section 2 we will give a brief overview of the current model for fuzzy regions 2.1.1, the first extension that makes use of the powerset (in 2.1.2) and its limitations (in 2.2). Section 3 concerns the proposed extension to use the fuzzy powerset in order to enrich the fuzzy regions and the resulting interpretation (in 3.3). The conclusion (section 4) summarizes the finding and mentions future work.

2 Preliminaries

In this section, the current definition for fuzzy regions and its extension using the powerset of the two dimensional domain is presented.

2.1 Fuzzy Regions

Traditionally, a polygon or other closed line is defined to be the boundary of a crisp region; the region is then considered to be inside this boundary ([6]). However, it is also possible to consider the region as the set of points contained in this boundary, and this is the view from which the fuzzy region can be defined.

2.1.1 Definition

From this point of view, it is a small step to augment the definition to a fuzzy set ([13], [14]) of points. In [8], the fuzzy region was defined over $\mathbb{R}^2$, thus with each element (point) a membership grade was associated.

Definition 1 (Fuzzy region)

$$\tilde{R} = \{(p, \mu_{\tilde{R}}(p)) | p \in \mathbb{R}^2\} \tag{1}$$

A fuzzy region essentially is a fuzzy set defined over a two dimensional domain; the concept is illustrated on figure 1. Consequently, the traditional fuzzy operations for intersection and union (t-norms and t-conorms) are immediately applicable. Specific spatial functionality has been added, some examples include the distance between regions and the (fuzzy) surface area of a region [10]. Topology has also been considered [11]; unlike in the crisp case appropriate definitions for the boundary, interior and exterior had to be derived from the initial given fuzzy set. The research is also still ongoing, as we try to find more optimal definitions. This model serves as a theoretical basis, in [7] and [9] we presented models suitable for implementation.

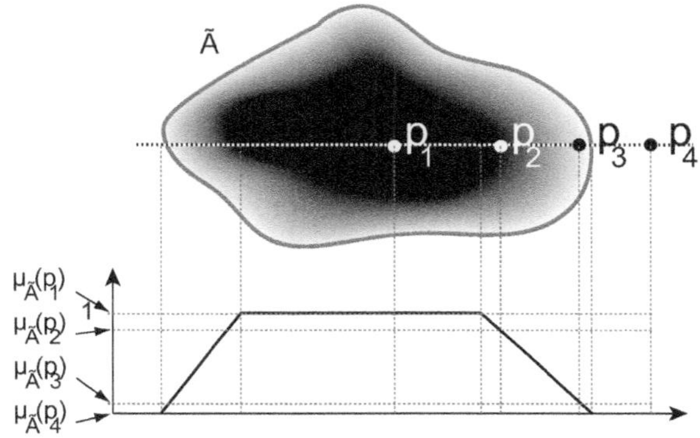

Fig. 1 The concept of a fuzzy region $\tilde{A}$; a fuzzy set over a two dimensional domain. All points belong to some extent to the region; indicated by means of the membership grade. The lower half of the figure shows a cross section. The shades of grey relate to the membership grades: darker shades match higher membership grades (the region has a dark outline to indicate its maximal outline).

The representation model can be used with a veristic interpretation to yield a fuzzy region: in this interpretation, the membership grades represent the extent to which the points belong to the region; but all points belong to the region. Giving the membership grades a possibilistic interpretation results in the representation of a fuzzy point: we are modelling a crisp point, and every element of fuzzy region is a candidate with its membership grade indicating the possibility. It goes without saying that the operators (e.g. distance) are impacted by this. In either interpretation, all points are considered to be independent of one another. In some situations, a user has added knowledge about the fact some points are linked (e.g. the example of the lake). A first extension to overcome this, makes use of the concept of the powerset.

2.1.2 Powerset Extension

To overcome the problem that all points are considered independently, elements of the fuzzy region need to be grouped. For this we first look at the concept: the fuzzy region is a fuzzy set of $\mathbb{R}^2$. To define the extension, we need to redefine the domain, and for this the powerset of $\mathbb{R}^2$ is used. The powerset $\wp$ of a set is a new set containing all the possible subsets of that particular set. To illustrate this, consider the following example:

$$\wp(\{0,1,2\}) = \{\{\},\{0\},\{1\},\{2\},\{0,1\},\{0,2\},\{1,2\},\{0,1,2\}\} \tag{2}$$

For $\mathbb{R}^2$ this becomes:

$$\wp(\mathbb{R}^2) = \{X | X \subseteq \mathbb{R}^2\} \tag{3}$$

Definition 2 shows the definition for fuzzy regions using the above powerset, as defined in [12].

Definition 2 (Fuzzy region with powerset extension)

$$\tilde{R} = \{(P, \mu_{\tilde{R}}(P)) | P \in \wp(\mathbb{R}^2) \wedge \forall P_1, P_2 \in \tilde{R} : P_1 \cap P_2 = \emptyset\} \quad (4)$$

Note that the intersection between any two elements should be empty: it is required that no two elements of the fuzzy region share points. A point can only be considered to belong to the region once, even if it is to a membership grade less than 1. When a fuzzy region is defined by means of a limited number of *subregions*, the concept bears resemblance to the concept of plateau regions [3]. The operations distance and surface have been considered in [12]. The extended concept of fuzzy regions is illustrated on figure 2, we refer to [12] for more details on the surface area. Simply put, the fuzzy surface area represents every possible surface area; this implies that when points are grouped in a subregion (e.g. the regions $\tilde{B}$ and $\tilde{C}$, there are less possibilities: the subregions either count as a whole, or do not count at all.

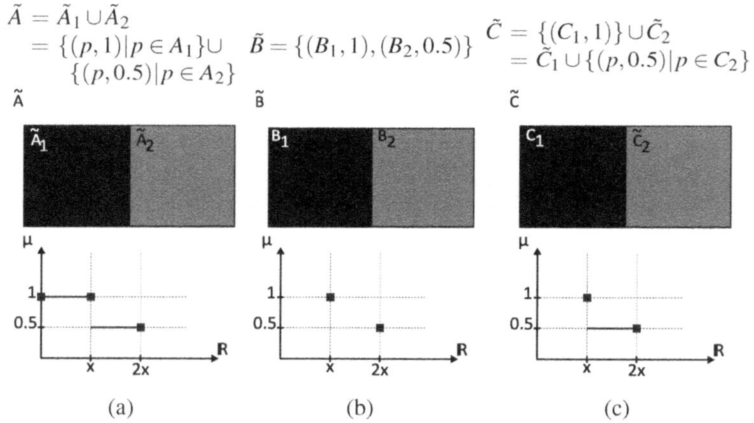

Fig. 2 Three different examples of fuzzy regions, with their surface areas: (a) the classical model, where each points is treated independently, (b) the extension with the powerset, showing a region consisting of two subregions each counted as a whole, (c) a region where there is both a subregion where points are treated independently, and a subregion that is counted as a whole. For each region, a mathematical explanation of its elements, a graphical illustration, where the shade of grey is representative of the membership grade, and its fuzzy surface area are shown.

2.1.3 Example

The fuzzy region based on the powerset still has the same interpretations as before. The lake from the example can now be represented as a fuzzy region, but with all the points that are at the same altitude (and thus would flood at the same time) contained within a subregion. While at first sight this makes no real difference to the representation of the region, this change in definition impacts its use. The fuzzy surface area [12] of the lake for instance will now be represented more accurately, possibly having less possible values than before, as points at the same altitude are not counted individually but as a group.

2.2 Limitations

While powerset extension mentioned in 2.1.2 allows for a richer model, it still is not without its limitations.

The first limitation is that all points contained within a subregion are required to have the same membership grade. This is basically caused by the fact that the subregion is a crisp region, which is given a membership grade as a whole. Additional knowledge could be present so that it may be necessary to group points with different membership grades.

The second limitation is more subtle and concerns the interpretations. Consider the example of the lake: it is said that the membership grades carry a veristic interpretation; all the points or subregions in the case of the powerset extension belong to the region to some extent. However, when we consider the changing water level on the lake, we could consider it to have a possibilistic interpretation, as there is only one water level at a time (it is just unknown to us for some reason). On the other hand, it is not possible to give the fuzzy region the possibilistic interpretation, as this would not represent a region anymore but more a fuzzy point (and a fuzzy set of candidate locations).

To overcome both limitations, an improved version of the powerset extension is presented.

3 Fuzzy Powerset Extension

3.1 Concept

The limitations of the previous extension were mentioned earlier. The first limitation, lack of grouping together points with different membership grades, could be solved by using fuzzy subregions. The use of fuzzy subregions will however introduce a second membership grade for the points of the region, which could serve as a solution to the interpretation problem.

The concept of the fuzzy powerset extension is similar to the previous extension: a fuzzy region will now be defined as a fuzzy set of fuzzy sets, which is achieved using the fuzzy powerset. The fuzzy powerset $\tilde{\wp}$ of a set A is the set of of all fuzzy sets over the given set A.

$$\tilde{\wp}(A) = \left\{\tilde{X} | \forall x : \mu_{\tilde{X}}(x) > 0 \Rightarrow x \in A\right\} \tag{5}$$

By using the $\tilde{\wp}(\mathbb{R}^2)$ as the domain for the fuzzy region, a region with fuzzy subregions can be defined.

3.2 Definition

Using the fuzzy powerset, it is possible to define a fuzzy region similarly as has been done with the powerset.

Definition 3 (Level-2 fuzzy region)

$$\tilde{R} = \{(\tilde{R}', \mu_{\tilde{R}}(\tilde{R}')) | \tilde{R}' \in \tilde{\wp}(\mathbb{R}^2)\} \tag{6}$$

The membership function is defined as:

$$\mu_{\tilde{R}} : \tilde{\wp}(\mathbb{R}^2) \mapsto [0,1]$$
$$\tilde{R}' \to \mu_{\tilde{R}}(\tilde{R}')$$

The elements of the fuzzy region $\tilde{R}$ are fuzzy regions as per definition 1; an important difference with the previous definition is that we now allow different subregions to share elements. The definition comprises what is referred to as a level-2 fuzzy set: a fuzzy set defined over a fuzzy domain ([2],[4]) and is named accordingly. This concept is not to be confused with a type-2 fuzzy set ([5]), which is a fuzzy set defined over a crisp domain but where the membership grades are fuzzy sets.

On figure 3, a region $\tilde{R}$ defined with three fuzzy subregions is shown; $\tilde{R} = \{(\tilde{R}'_1, \mu_{\tilde{R}}(\tilde{R}'_1)), (\tilde{R}'_2, \mu_{\tilde{R}}(\tilde{R}'_2)), (\tilde{R}'_3, \mu_{\tilde{R}}(\tilde{R}'_3))\}$.

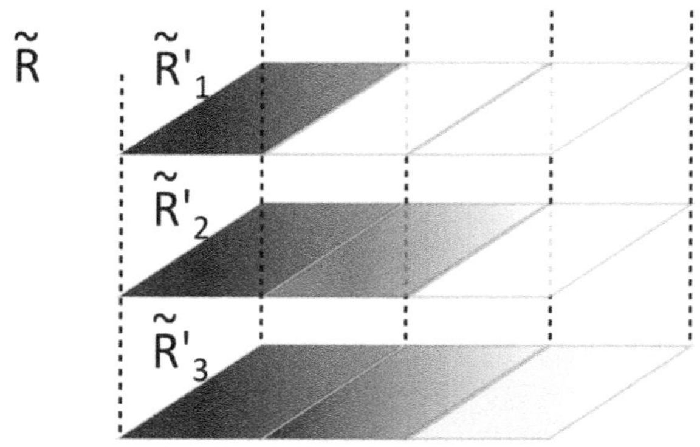

Fig. 3 A fuzzy region as $\tilde{R}$ defined using definition 3. $\tilde{R}$ has three overlapping subregions ($\tilde{R}'_1$, $\tilde{R}'_2$ and $\tilde{R}'_3$); each fuzzy region (1), as indicated by the grey scales. These subregions are candidate representations for the feature modelled, and carry membership grades to indicate this possibility (not shown). As before, darker colours indicate higher membership grades.

3.3 Interpretation

By using fuzzy regions (definition 1) as basic elements for the new concept, points of the universe will have multiple membership grades associated: some from their own membership in the subregion, some from the membership of the subregion that contains them. To explain the interpretations, first the new definition will be considered under the same limitation as the powerset extension: subregions will not be allowed to overlap. This simplification has quite an impact on the interpretation, but makes the link with the previous extension more obvious.

3.3.1 Non Overlapping Subregions

In the first extension (section 2.1.2), subregions were not allowed to overlap in order to simplify the model. In this section, we will consider the new extension (section 3) under the same limitation, merely to illustrate the way it impacts the interpretation. By limiting the definition in not allowing overlapping subregions, it becomes a straight forward extension of the previous definition (2.1.2): the fuzzy set over the new domain (i.e. the region $\tilde{R}$) needs to have a veristic interpretation as it is represents a region (all elements belong to it). The fuzzy set in each subregion $\tilde{R}'$ also needs to carry a veristic interpretation (or the subregion would not be a correct representation of a region). An example of such a region is shown on figure 4.

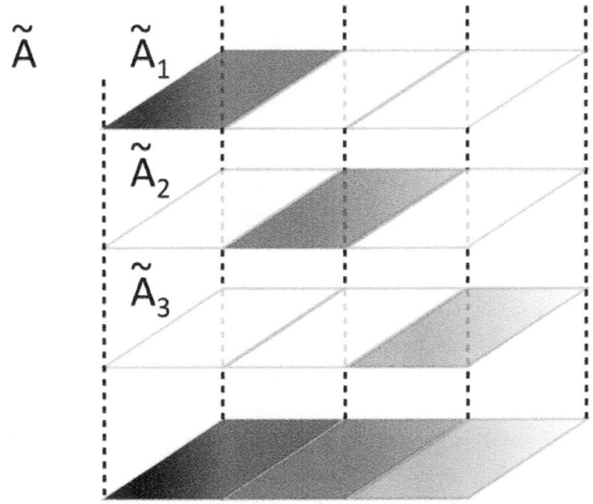

Fig. 4 Illustration of a fuzzy region as $\tilde{A}$ defined using the above definition 3, but not allowing regions to overlap. It shows three subregions ($\tilde{A}'_1$, $\tilde{A}'_2$ and $\tilde{A}'_3$). Each of these subregions is a fuzzy region (using definition 1) on its own, as indicated by the grey scales. The feature being modelled is the union of all these subregions; and therefore contains each of them to some extent. Each subregion carries a membership grade to indicate this extent. As before, darker colours indicate higher membership grades.

Points p of the universe $\mathbb{R}^2$ carry only two membership grades, as they only belong to one $\tilde{R}' \in \tilde{R}$: the membership grade $\mu_{\tilde{(R)}}(R')$ and indirectly $\mu_{\tilde{(R')}}(p)$. To say something about p in relation to $\tilde{R}$, these grades can be accumulated, using the intersection of the membership grades of the point in the subregion, and the membership grade of the subregion. This allows us to define the function $\mu'_{\tilde{R}}$ which returns the membership for individual points of the universe (it is not a true membership function, we will elaborate on this in the next section).

Despite the presence of the limitation, the first problem described in 2.2 is solved: points with different membership grades can be grouped in one subregion. The second problem however, the ambiguity regarding the interpretation, remains.

Example

To simplify the example, the regions will be defined as finite sets of points. The formulas also hold for infinite sets. Consider the region $\tilde{R}$ defined as below.

$$\tilde{R} = \{(\tilde{R'_1}, 1), (\tilde{R'_2}, 0.5)\} \text{ with } \begin{cases} \tilde{R'_1} = \{(p_1^1, 0.8), (p_2^1, 0.6)\} \\ \tilde{R'_2} = \{(p_1^2, 0.8), (p_2^2, 0.4)\} \end{cases}$$

Using a t-norm T, the membership grades μ' are

$$\begin{aligned} \mu'_{\tilde{R}}(p_1^1) &= T(\mu_{\tilde{R}}(\tilde{R'_1}), \mu_{\tilde{R'_1}}(p_1^1)) = T(1, 0.8) \\ \mu'_{\tilde{R}}(p_2^1) &= T(\mu_{\tilde{R}}(\tilde{R'_1}), \mu_{\tilde{R'_1}}(p_2^1)) = T(1, 0.6) \\ \mu'_{\tilde{R}}(p_1^2) &= T(\mu_{\tilde{R}}(\tilde{R'_2}), \mu_{\tilde{R'_1}}(p_1^2)) = T(0.5, 0.8) \\ \mu'_{\tilde{R}}(p_2^2) &= T(\mu_{\tilde{R}}(\tilde{R'_2}), \mu_{\tilde{R'_1}}(p_2^2)) = T(0.5, 0.4) \end{aligned}$$

3.3.2 Overlapping Subregions

For regions with overlapping subregions, the situation becomes more complex. Each of the subregions $\tilde{R}'$ can be seen as a *candidate* fuzzy region, but as a region has a veristic interpretation of its elements. Each subregion $\tilde{R}'$ is a basic element of a fuzzy set $\tilde{R}$, in which it will carry a membership grade $\mu_{\tilde{R}}(\tilde{R}')$ to indicate its possibility, and as such a possibilistic interpretation is needed on this second level. The model can thus be used to represent fuzzy points as well: it suffices to consider singleton sets subregions. We can now consider the relationship between a single point p of the universe $\mathbb{R}^2$ with the fuzzy region $\tilde{R}$. While it is of course not possible to consider a point of a subregion $\tilde{R}'$ independently from other points of the same subregion $\tilde{R}'$, it is possible to provide information regarding individual points. A point of the universe, if it is contained within more than one subregion, will have a number of membership grades from the first level and a number of membership grades from the second level. The former relate to the extent to which the point belongs to the region, whereas the latter relate to the possibility the point belongs to the region. Expressing the membership of this point in relation to the fuzzy region $\tilde{R}$ implies there is uncertainty about the points membership grade. To express this, we can resort to a fuzzy set to describe the membership.

Definition 4 (Membership of points)

$$\begin{aligned} \mu'_{\tilde{R}}(p) : \mathbb{R}^2 &\mapsto \widetilde{[0,1]} \\ p &\rightarrow \mu_{\tilde{R}}(p) \end{aligned}$$

The membership function is defined as

$$\mu'_{\tilde{R}}(p) = \bigcup_{\tilde{R}' \in \tilde{R}} \{(\mu'_{\tilde{R}'}(p), \mu_{\tilde{R}}(\tilde{R}'))\} \tag{7}$$

Note that $\mu'_{\tilde{R}}(p)$ is not a true membership function, as $\tilde{R}$ is not a set of points. The function is also not normalized: a point can belong to an extent smaller than 1 to regions that have a possibility smaller than 1. From this definition, it can be seen that while the region was defined as level-2 fuzzy set (fuzzy set over $\tilde{\wp}(\mathbb{R}^2)$); we can consider it a type-2 fuzzy set (over $\mathbb{R}^2$), but in this latter view we loose the knowledge about the subregions. The value of this function is that it allows us to make statements on individual points of the universe in relation to a fuzzy region.

Example

To represent the region, all the outlines for the possible boundary need to be considered, along with how possible each outline is (some outlines may be quite likely, whereas others are not). The outline can have points with different membership grades to indicate an imprecise outline. A simple example using finite sets to illustrate the functions is given below. Each fuzzy membership has a possibilistic

$$\tilde{R} = \{(\tilde{R}'_1, 1), (\tilde{R}'_2, 0.6), (\tilde{R}'_3, 0.4)\} \text{ with } \begin{cases} \tilde{R}'_1 = \{(p_1, 1), (p_2, 0.7)\} \\ \tilde{R}'_2 = \{(p_1, 0.8), (p_2, 0.7)\} \\ \tilde{R}'_3 = \{(p_1, 0.4), (p_3, 0.8)\} \end{cases}$$

The membership grades for the three points then are:

$$\begin{aligned} \mu'_{\tilde{R}}(p_1) &= \{(\mu'_{\tilde{R}'_1}(p_1), \mu'_{\tilde{R}}(\tilde{R}'_1))\} \cup \{(\mu'_{\tilde{R}'_2}(p_1), \mu'_{\tilde{R}}(\tilde{R}'_2)\} \cup \{(\mu'_{\tilde{R}'_3}(p_1), \mu'_{\tilde{R}}(\tilde{R}'_3))\} \\ &= \{(1,1), (0.8, 0.6), (0.4, 0.4)\} \\ \mu'_{\tilde{R}}(p_2) &= \{(\mu'_{\tilde{R}'_1}(p_2), \mu'_{\tilde{R}}(\tilde{R}'_1))\} \cup \{(\mu'_{\tilde{R}'_2}(p_2), \mu'_{\tilde{R}}(\tilde{R}'_2))\} = \{(0.7, S(1, 0.6))\} = \{(0.7, 1)\} \\ \mu'_{\tilde{R}}(p_3) &= \{(\mu'_{\tilde{R}'_3}(p_3), \mu'_{\tilde{R}}(\tilde{R}'_3))\} = \{(0.8, 0.4)\} \end{aligned}$$

interpretation to indicate the possibility the element belongs to the region to the given extent. The point p_1 belongs to an extent 1 with a possibility of 1 (if it is in subregion $\tilde{R}'_1$), to an extent 0.8 with possibility 0.6 (subregion $\tilde{R}'_2$) and to an extent 0.4 with possibility 0.4 (subregion $\tilde{R}'_3$). For p_2 an s-norm is needed as it belongs to the same extent to two regions; in this example the maximum was used.

4 Conclusion

In this contribution, we presented an important extension to our model for fuzzy regions, yielding level-2 fuzzy regions. The extension allows for two levels of uncertainty or imprecision, allowing the representation of features that partly belong to the region, and features that possibly belong to the region (or any combination). This change makes for a much richer modelling, allowing the fuzzy regions to represent real life features more closely. The extension also unifies the representation of fuzzy regions and fuzzy points, overcoming the need of specifying the interpretation. Obviously the changes require adaptation of a number of operations that have been defined so far, which is our main focus of the future work.

References

1. Dubois, D., Prade, H.: Fundamentals of Fuzzy Sets. Kluwer Academic Publishers, Dordrecht (2000)
2. Gottwald, S.: Set theory for fuzzy sets of higher level. Fuzzy Sets and Systems 2(2), 125–151 (1979)
3. Kanjilal, V., Liu, H., Schneider, M.: Plateau regions: An implementation concept for fuzzy regions in spatial databases and GIS. In: Hüllermeier, E., Kruse, R., Hoffmann, F. (eds.) IPMU 2010. LNCS, vol. 6178, pp. 624–633. Springer, Heidelberg (2010)
4. Klir, G.J., Yuan, B.: Fuzzy sets and fuzzy logic: Theory and applications. Prentice Hall, New Jersey (1995)
5. Mendel, J.M.: Uncertain Rule-Based Fuzzy Logic Systems, Introduction and New Directions. Prenctice Hall PTR, Englewood Cliffs (2001)
6. Rigaux, P., Scholl, M., Voisard, A.: Spatial Databases with Applications to GIS. Morgan Kaufman Publishers, San Francisco (2002)
7. Verstraete, J., De Tré, G., Hallez, A.: Adapting TIN-layers to Represent Fuzzy Geographic Information. In: The 7th Meeting of the EURO Working Group on Fuzzy Sets, pp. 57–62 (2002)
8. Verstraete, J., De Tré, G., De Caluwe, R., Hallez, A.: Field Based Methods for the modelling of Fuzzy Spatial Data. In: Petry, F., Robinson, V., Cobb, M. (eds.) Fuzzy modeling with Spatial Information for Geographic Problems, pp. 41–69. Springer, Heidelberg (2005)
9. Verstraete, J., Hallez, A., De Tré, G.: Bitmap Based Structures for the modelling of Fuzzy Entities. Special issue of Control & Cybernetics 35(1), 147–164 (2006)
10. Verstraete, J.: Fuzzy regions: interpretations of surface area and distance. Control and Cybernetics 38, 509–528 (2009)
11. Verstraete, J.: A quantitative approach to topology for fuzzy regions. In: Rutkowski, L., Scherer, R., Tadeusiewicz, R., Zadeh, L.A., Zurada, J.M. (eds.) ICAISC 2010. LNCS (LNAI), vol. 6113, pp. 248–255. Springer, Heidelberg (2010)
12. Verstraete, J.: Fuzzy regions: adding subregions and the impact on surface and distance calculation. In: Hüllermeier, E., Kruse, R., Hoffmann, F. (eds.) Communications in Computer and Information Science, Part 1, vol. 80, pp. 561–570 (2010)
13. Zadeh, L.A.: Fuzzy Sets. Information and Control 1(3), 338–353 (1965)
14. Zimmerman, H.-J.: Practical Applications of Fuzzy Technologies. Kluwer Academic Publishers, Dordrecht (1999)

Evaluation of Categorical Data Clustering

Hana Rezankova, Tomas Loster, and Dusan Husek

Abstract. Methods of cluster analysis are well known techniques of multivariate analysis used for many years. Their main applications concern clustering objects characterized by quantitative variables. For this case various coefficients for clustering evaluation and determination of cluster numbers have been proposed. However, in some areas, i.e., for segmentation of Internet users, the variables are often nominal or ordinal as their origin in questionnaire responses. That is why we are dealing with the evaluation criteria for the case of categorical variables here. The criteria based on variability measures are proposed. Instead of variance as a measure for quantitative variables, three measures for nominal variables are considered: the variability measure based on a modal frequency, Gini's coefficient of mutability, and the entropy. The proposed evaluation criteria are applied to a real-dataset.

Keywords: Cluster analysis, Nominal variable, Determination of cluster numbers, Evaluation of clustering.

1 Introduction

Cluster analysis is an important tool in many areas: for example, for segmentation of Internet users. This segmentation can be based on a questionnaire when responses are on the nominal and ordinal scales. In [17] groups of

Hana Rezankova · Tomas Loster
University of Economics, Prague, nam. W. Churchilla 4,
13067 Praha 3, Czech Republic
e-mail: hana.rezankova@vse.cz, tomas.loster@vse.cz

Dusan Husek
Institute of Computer Science,
Academy of Sciences of the Czech Republic,
Pod vodarenskou vezi 2, 18207 Praha 8, Czech Republic
e-mail: dusan@cs.cas.cz

E. Mugellini et al. (Eds.): Advances in Intelligent Web Mastering - 3, AISC 86, pp. 173–182.
springerlink.com

Internet users based on their relation to advertisement are identified. These groups are characterized from the sociodemographic aspect. If the sociodemographic users profile of portals is analyzed, specific recommendations for the advertisement submitters can be prepared.

Many cluster analysis methods have been proposed, including special techniques for categorical data, see [1, 3, 5, 9, 10, 11, 12]. Further, many coefficients for clustering evaluation and determination of cluster numbers have been proposed. However, these coefficients serve mainly for the case when objects are characterized by quantitative variables [4, 8, 13]. That is why we propose new evaluation criteria for the case when objects are characterized by nominal variables.

The simplest way to cluster objects characterized by nominal variables in statistical software packages is a proximity matrix creation using the coefficient of disagreement, followed by hierarchical cluster analysis. The *coefficient of disagreement* is calculated as a ratio of the number of variables with distinct values for pairs of the objects and the total number of variables.

Another measure of the relationship between two objects (and between two clusters as well) is the *log-likelihood distance*. Its implementation in the software products is tied with *two-step cluster analysis* in the SPSS system (now IBM SPSS Statistics). This procedure, based on the BIRCH algorithm [18], has been designed for clustering of a large number of objects. The log-likelihood distance is determined for data files combining quantitative and nominal variables. Dissimilarity is expressed on the basis of variability and the entropy is applied to nominal variables as the variability measure. For the lth variable in the gth cluster, the *entropy* is calculated according to the formula

$$H_{gl} = -\sum_{u=1}^{K_l} \left(\frac{n_{glu}}{n_g} \ln \frac{n_{glu}}{n_g} \right), \tag{1}$$

where K_l is the number of categories of the lth variable, n_{glu} represents the frequency of the uth category of the lth variable in the gth cluster, and n_g is the number of objects in the gth cluster. The minimum of this variability measure is 0 and the maximum is $\ln K_l$. If objects are characterized by only nominal variables, we can calculate the entropy of a certain cluster as a sum of the entropies for all variables. Two objects are the most similar if the cluster composed of them has the smallest entropy.

The within-cluster entropy of the data set divided to k clusters can be written as

$$H(k) = \sum_{g=1}^{k} \frac{n_g}{n} \sum_{l=1}^{m} H_{gl}, \tag{2}$$

where k is the number of clusters, m is the number of nominal variables, and n is the number of objects. Values calculated on the basis of results of different clustering methods can be used for evaluation of these methods. The smaller values indicate smaller variability and hence, better clustering. The within-cluster entropy is a basis for the COOLCAT algorithm [1].

As regards of determination of a cluster number, an information criterion can be used as a basis. In SPSS, both *Schwarz's Bayesian information criterion* (BIC) and *Akaike's information criterion* (AIC) are implemented. For the case when all variables are nominal, the former is calculated as

$$I_{BIC}(k) = 2n \cdot H(k) + k \left(\sum_{l=1}^{m} (K_l - 1) \right) \ln(n), \tag{3}$$

the latter is calculated as

$$I_{AIC}(k) = 2n \cdot H(k) + 2k \left(\sum_{l=1}^{m} (K_l - 1) \right). \tag{4}$$

Moreover, latent class models [14] make mixed type data clustering possible, including determination of cluster numbers. This approach is implemented in the Latent GOLD software. It is also possible to transform a data file with nominal variables to a data file with binary variables and use techniques of cluster analysis for binary or quantitative data.

This paper is organised as follows. In Sect. 2 we describe variability measures for nominal variables and we propose variability measures based evaluation criteria of clustering. In Sect. 3 the results of criteria application to a real-data file are compared. The conclusion includes characterizing of these criteria.

2 Evaluation Criteria of Clustering

Quality of partitioning of objects to clusters can be evaluated in different ways. The aim of the disjunctive clustering is to create clusters with a small variability of variables within them. In the case of quantitative variables, the variance is applied as a variability measure. The variability within clusters, between clusters and the total variability of all variables are investigated. It is an analogy of one factor MANOVA (multivariate analysis of variance) where a new variable containing a label of clusters to which the object was assigned is a factor.

2.1 Variability Measures for Nominal Variables

In the previous section, entropy (1) as a measure of variability for nominal variables was mentioned. Two other measures and their normalized forms can be used for this purpose. The first is to only consider the relative frequency of the modal category, i.e., for the lth variable in the gth cluster the *modal frequency based variability measure* is calculated as

$$V_{gl} = 1 - \frac{\max_{u=1,2,\ldots,K_l}\{n_{glu}\}}{n_g}. \tag{5}$$

The minimum of this measure is 0 and the maximum is $(K_l - 1)/K_l$. If this measure is divided by the maximum possible value, we obtain the normalized measure V'_{gl} with the values from the interval [0, 1].

The other variability measure is *Gini's coefficient* (a measure of mutability, see [6]) which is calculated as

$$G_{gl} = 1 - \sum_{u=1}^{K_l} \left(\frac{n_{glu}}{n_g} \right)^2 . \tag{6}$$

The minimum of this measure is 0 and the maximum is $(K_l - 1)/K_l$, as well as in the previous case. By dividing by the maximum possible value we obtain the normalized measure G'_{gl} with the values from the interval [0, 1].

Similarly, if the entropy is divided by the maximum possible value ($\ln K_l$), we obtain the normalized measure H'_{gl} with the values from the interval [0, 1].

2.2 Variability Measures Based Evaluation Criteria of Clustering

If an analogy of MANOVA is applied, several coefficients have been proposed for the quantitative data. All of these coefficients can be modified for nominal and ordinal variables. In the following text, we propose the applications of variability measures (1), (5), and (6).

We can distinguish a comparison of results of different clustering methods and determination of a suitable number of clusters. In the latter, the local optimum from the specified range is usually searched.

As for method comparison, we suppose that the partitioning results are compared for the same number of clusters. In this case the total variability (mutability or entropy) in the data set (the sum of the variability of the all variables) is the same, and so sufficient criterion for comparison is the evaluation of the within-cluster variability only.

The *within-cluster variability* for k clusters based on the modal frequency with using the formula (5) is calculated as

$$V(k) = \sum_{g=1}^{k} \frac{n_g}{n} \sum_{l=1}^{m} V_{gl} . \tag{7}$$

The smaller values indicate smaller variability and hence, better clustering. We obtain the value from the interval [0, 1] with the use of the normalized measure:

$$V'(k) = \sum_{g=1}^{k} \frac{n_g}{n \cdot m} \sum_{l=1}^{m} \frac{K_l}{K_l - 1} V_{gl} . \tag{8}$$

For determination of a suitable cluster number, the differences between the values of within-cluster variability can be calculated. However, some other approaches are usually applied, see below. If we consider an analogy with the R-squared coefficient from ANOVA, the ratio of the between-cluster variability and the total variability – index $I_{RSQ}(k)$ – can be calculated. We also obtain the values from the interval [0, 1]. In this case, however, the greater value indicates the better clustering.

The modification of R-square with the use of the *modal frequency based variability measure* for one dependent nominal variable is called the *Goodman and Kruskal's lambda* (an association measure). Using $V(k)$ for m nominal variables and k clusters we can define *lambda index*

$$I_\lambda(k) = \frac{V(1) - V(k)}{V(1)}, \tag{9}$$

where $V(1)$ is the variability of the whole data set ($k = 1$). In this case the value of lambda index is 0.

The *within-cluster mutability* for k clusters $G(k)$ based on Gini's coefficient (6) is computed analogously as the within-cluster variability based on the modal frequency, see the formula (7), as well as the normalized form $G'(k)$, see the formula (8). The modification of R-square with the use of the *Gini's coefficient* as a variability measure for one dependent nominal variable is called the *Goodman and Kruskal's tau*, see [7]. It is an association measure for two nominal variables. Using $G(k)$ for m nominal variables we can define *tau index* $I_\tau(k)$, analogously to the formula (9) in wich $V(k)$ is replaced by $G(k)$ and $V(1)$ is replaced by $G(1)$.

As concern as the *within-cluster entropy* (2), we obtain the value from the interval [0, 1] with the use of the normalized entropy:

$$H'(k) = \sum_{g=1}^{k} \frac{n_g}{n \cdot m} \sum_{l=1}^{m} \frac{H_{gl}}{\ln K_l}. \tag{10}$$

The modification of R-square with the use of the entropy as a variability measure for one dependent nominal variable is called the *coefficient of uncertainty* [15]. It is an association measure for two nominal variables. Using $H(k)$ defined in (2) for m nominal variables we can define *uncertainty index* $I_U(k)$, analogously to the formula (9) in wich $V(k)$ is replaced by $H(k)$ and $V(1)$ is replaced by $H(1)$.

For all measures mentioned above we can see that partitioning to a greater number of clusters results in lower within-cluster variability, and a ratio of between-cluster variability and the total variability achieves the greater values. For this reason, for determination of cluster numbers some special indices were proposed. There are for example semipartial R-squared index (SPRSQ) and pseudo F (PSF, CHF) index.

The semipartial R-squared index for k clusters (implemented in the SAS system) is expressed as

$$I_{SPRSQ}(k) = I_{RSQ}(k+1) - I_{RSQ}(k). \tag{11}$$

For nominal variables, we propose to use $I_\lambda(k)$, $I_\tau(k)$ or $I_U(k)$ instead $I_{RSQ}(k)$ and calculate the semipartial indices $I_{SP\lambda}(k)$, $I_{SP\tau}(k)$, and $I_{SPU}(k)$. The smallest value indicates the best partitioning of objects into clusters.

The CHF (Calinski and Habarasz) index (pseudo F index) is the ratio of between-cluster variance and within-cluster variance [2]. The formula of the modified index based on the modal frequency is

$$I_{PSF_\lambda}(k) = \frac{(n-k)(V(1)-V(k))}{(k-1)V(k)}. \qquad (12)$$

Similarly, we can calculate indices $I_{PSF\tau}(k)$ based on the Gini's coefficient and $I_{PSFU}(k)$ based on the entropy. If the data set is not divided to clusters, the between-cluster variability is zero and also the value of each mentioned index is 0.

3 Example

We analyzed the data file created on the basis of the answers provided by 50 participants of the seminar. We chose the variables with frequencies of at least 10 for each category. For this reason, some new variables were created both by merging variables and by joining categories. The analyzed data set contains seven categorical variables (for all pairs, the variables are independent at 5% significance level when chi-square test for independence is used). There are four binary variables (interest on news in research, participant of either the mathematics or physics Olympiad, participant of biology Olympiad, and participant of some other Olympiad), two variables with three categories (where the respondent got information about the seminar with categories of school, educational camp, and other place; and class of the secondary school with categories of first and second, third, and fourth), and the dichotomous variable of sex (there were 25 males and 25 females). For analyses, we used the SPSS Statistics system.

For illustration of categorical data analysis, we applied two methods available in statistical packages: two-step cluster analysis with the log-likelihood distance (TS) and hierarchical cluster analysis with the complete linkage method (CL). We used SPSS but we prepared the proximity matrix based on the coefficient of disagreement in STATISTICA. Minor reordering of the dataset was performed before the analyses (the first experiment was done for all nominal variables and original categories, hierarchical cluster analysis was applied; the rows of the data matrix were assigned to six clusters and sorted according to them). We clustered participants of the seminar, considering from two to six clusters.

For two clusters the sizes of the individual clusters were 33 and 17 for the TS method and 23 and 27 for the CL method. For 6 clusters the sizes were from 4 to 14. We calculated variability measures, other evaluation criteria mentioned in Chapter 2, and information criteria BIC and AIC. Tables

1 and 2 contain the values of these measures for the individual numbers of clusters both for hierarchical cluster analysis with the complete linkage method (Table 1) and for two-step cluster analysis (Table 2).

Within-cluster variability measured by both Gini's coefficient, $G'(k)$, and the entropy, $H'(k)$, is smaller for TS clustering for all numbers of clusters. By analogy, tau index $I_{\tau}(k)$, and uncertainty index $I_U(k)$ give the higher values for TS clustering. That means that the partitioning of objects into clusters is better using TS clustering in the described example. The third mentioned

Table 1 Evaluation of the results obtained by complete linkage method

Measure	Number of clusters					
	1	2	3	4	5	6
$V'(k)$	0.71	0.58	0.49	0.45	0.41	0.39
$G'(k)$	0.87	0.75	0.66	0.62	0.57	0.53
$H'(k)$	0.90	0.79	0.70	0.65	0.59	0.56
$I_{\lambda}(k)$	0	0.17	0.30	0.36	0.42	0.46
$I_{\tau}(k)$	0	0.14	0.23	0.29	0.35	0.39
$I_U(k)$	0	0.12	0.20	0.28	0.35	0.39
$I_{SP\lambda}(k)$	0.17	0.13	0.06	0.06	0.04	–
$I_{SP\tau}(k)$	0.14	0.10	0.06	0.06	0.04	–
$I_{SPU}(k)$	0.12	0.09	0.08	0.07	0.04	–
$I_{PSF\lambda}(k)$	0	10.11	**10.28**	8.71	8.16	7.39
$I_{PSF\tau}(k)$	0	**7.49**	7.11	6.33	6.08	5.63
$I_{PSFU}(k)$	0	**6.24**	5.94	5.89	6.01	5.66
$I_{BIC}(k)$	548.39	524.60	515.23	511.22	**510.64**	523.64
$I_{AIC}(k)$	531.18	490.18	463.60	442.38	424.60	420.39

Table 2 Evaluation of the results obtained by two-step cluster analysis

Measure	Number of clusters					
	1	2	3	4	5	6
$V'(k)$	0.71	0.55	0.47	0.47	0.43	0.38
$G'(k)$	0.87	0.73	0.62	0.58	0.53	0.48
$H'(k)$	0.90	0.77	0.65	0.59	0.54	0.48
$I_{\lambda}(k)$	0	0.22	0.33	0.34	0.40	0.46
$I_{\tau}(k)$	0	0.15	0.28	0.32	0.39	0.45
$I_U(k)$	0	0.13	0.26	0.33	0.39	0.46
$I_{SP\lambda}(k)$	0.22	0.11	**0.01**	0.06	0.06	–
$I_{SP\tau}(k)$	0.15	0.12	**0.05**	0.06	0.07	–
$I_{SPU}(k)$	0.13	0.13	0.07	0.06	0.06	–
$I_{PSF\lambda}(k)$	0	**13.33**	11.37	7.92	7.46	7.39
$I_{PSF\tau}(k)$	0	8.74	**9.01**	7.33	7.11	7.29
$I_{PSFU}(k)$	0	7.32	**8.28**	7.61	7.30	7.39
$I_{BIC}(k)$	548.39	515.79	485.10	**483.85**	487.23	490.15
$I_{AIC}(k)$	531.18	481.38	433.47	415.01	401.19	386.90

variability measure considers only frequency of one category (modal) therefore it is less suitable for use, as is lambda index $I_\lambda(k)$.

Using modified semipartial coefficients (within the interval from 1 to 5 clusters), with semipartial lambda index $I_{PSF\lambda}(k)$ and semipartial tau index $I_{SP\tau}(k)$, three clusters were identified as optimal in the use of TS clustering. In this case the semipartial uncertainty index $I_{SPU}(k)$ it was four clusters (0.062 in comparison of 0.063 for 5 clusters. Using the CL method, the smallest values using all three coefficients are for 5 clusters but it is not known if there is a local minimum. Using modified pseudo F indices, we obtained 2 and 3 clusters as optimal.

For the purpose of comparison of proposed criteria with approaches implemented in software packages, we applied information criteria BIC and AIC. In the case of the CL method the smallest value of $I_{BIC}(k)$ is taken on 5 clusters. In the case of TS clustering, the smallest value of $I_{BIC}(k)$ is taken on 4 clusters. On the basis of the AIC, the smallest values are for 6 clusters but it is not known if there is a local minimum. In SPSS, the largest increase in distance between the two closest clusters in each hierarchical clustering stage is the final criterion for the determination of cluster numbers. In our case it was 4 clusters for TS clustering. For both criteria, the values are lower for clustering obtained by TS clustering.

We can summarize that the results obtained by TS clustering was evaluated better than results obtained by the CL method by almost all applied criteria for all clusters (modal frequency based criteria gave distinct evaluation for some clusters). As concerns determination of cluster numbers, the results differ. For TS clustering, 3 clusters were indicated in 4 cases, 4 clusters in 2 cases and 2 clusters in 1 case. For CL method it is not known if local minima were found in some cases, as well as in the case of the AIC criterion generally.

We can characterize 3 clusters obtained by the TS method by the following way. In the first cluster the respondents are only boys mostly from the first and second classes, who got information about seminar in the camp, are not interested in research news and do not participate in Olympiads. The second cluster is represented by boys and girls, who got information about seminar mostly at school. They are not interested in research news and they participate mostly in Biology or some other Olympiads. In the third cluster girls from the third class predominate, who got information about seminar both at school and the camp. They are mostly interested in research news.

In our presented example we evaluated two-step cluster analysis as the methods with better results in comparison to the hierarchical clustering by complete linkage method on the basis of coefficient of disagreement. However, the results of the former method are dependent on the object order. For a different order we obtain different assignments of objects to clusters (and a different optimal numbers of clusters).

The advantage of hierarchical cluster analysis is object order independence. The problem with application of this method can come when the data file is too large. Another problem is that the coefficient of disagreement is needed for clustering the objects characterized by nominal variables, what it is not always implemented, for example, in such a system as SPSS is. The user has to prepare the proximity matrix in another way.

4 Conclusion

In the paper, we proposed several evaluation criteria for categorical data clustering based on three different variability measures for nominal variables. Only one of them – entropy – is usually used for this purpose. It is applied in information criteria BIC and AIC. In addition, we use Gini's coefficient of mutability and the modal frequency based variability measure. Similarly, we could use the variability measure for ordinal variables based on cumulative frequencies of individual categories.

For comparison of results of different clustering methods the use of some within-cluster variability measure is sufficient. In case when the objects characterized by nominal variables are clustered, either Gini's coefficient of mutability or entropy can be used. The measurement of variability only on the basis of the modal frequency is less suitable. The normalized within-cluster variability measures and proposed lambda index, tau index and uncertainty index are useful if clustering evaluation by a value from the interval [0, 1] is needed.

As concerns determination of cluster numbers, three clusters were indicated as optimal by 4 from 6 proposed indices in the case of TS clustering whereas four clusters were indicated by the BIC criterion and the result obtained by the AIC criterion was not clear. According to our experiences with data clustering this criterion usually does not indicate the local minimum within the specified interval.

Our further research will focus on detail evaluation of the proposed criteria including criteria for a case when objects characterized by mixed type variables are clustered.

Acknowledgements. This work was supported by projects AV0Z10300504, GACR P202/10/0262, 205/09/1079, MSM6138439910, and IGA VSE F4/3/2010.

References

1. Barbará, D., Li, Y., Couto, J.: COOLCAT: An entropy-based algorithm for categorical clustering. In: Proceedings of the 11th International Conference on Information and Knowledge Management, pp. 582–589. ACM Press, McLean (2002)
2. Calinski, T., Habarasz, J.: A dendrite method for cluster analysis. Communications in Statistics 3, 1–27 (1974)
3. Chatuverdi, A., Foods, K., Green, P.E., Carroll, J.D.: K-modes clustering. Journal of Classification 18, 35–55 (2001)

4. Gan, G., Ma, C., Wu, J.: Data Clustering: Theory, Algorithms, and Applications. ASA-SIAM, Philadelphia (2007)
5. Ganti, V., Gehrke, J., Ramakrishnan, R.: CACTUS – Clustering categorical data using summaries. In: Proceedings of the 5th ACM SIGKDD International Conference on Knowledge Discovery and Data Mining, pp. 73–83. ACM Press, San Diego (1999)
6. Gini, C.W.: Variability and Mutability. Contribution to the study of statistical distributions and relations. Studi Economico-Giuridici della R. Universita de Cagliari (1912); Reviewed in: Light, R.J., Margolin, B.H.: An Analysis of Variance for Categorical Data. J. American Statistical Association 66, 534–544 (1971)
7. Goodman, L.A., Kruskal, W.H.: Measures of association for crossclassification. Journal of the American Statistical Association 49, 732–764 (1954)
8. Gordon, A.D.: Classification, 2nd edn. Chapman & Hall/CRC, Boca Raton (1999)
9. Guha, S., Rastogi, R., Shim, K.: ROCK: A robust clustering algorithm for categorical attributes. Information Systems 25, 345–366 (2000)
10. He, Z., Xu, X., Deng, S.: Squeezer: An efficient algorithm for clustering categorical data. Journal of Computer Science and Technology 17, 611–625 (2002)
11. Huang, Z.: A fast clustering algorithm to cluster very large categorical data sets in data mining. In: Proceedings of the SIGMOD Workshop on Research Issues on Data Mining and Knowledge Discovery, University of British Columbia, pp. 1–8 (1997)
12. Huang, Z.: Extensions to the k-means algorithm to clustering large data sets with categorical values. Data Mining and Knowledge Discovery 2, 283–304 (1998)
13. Kogan, J.: Introduction to Clustering Large and High-Dimensional Data. Cambridge University Press, New York (2007)
14. Magidson, J., Vermunt, J.K.: Latent class models for clustering: A comparison with K-means. Canadian Journal of Marketing Research 20, 37–44 (2002)
15. Press, W.H., Flannery, B.P., Teukolsky, S.A., Vetterling, W.T.: Numerical Recipes in C: The Art of Scientific Computing, p. 634. Cambridge University Press, Cambridge (1988)
16. Sharma, S.: Applied Multivariate Techniques. John Wiley & Sons, Inc., New York (1995)
17. Sila, M.: Analysis of Internet Visits and Internet Users (in Czech). Diploma thesis. University of Economics, Prague (2010)
18. Zhang, T., Ramakrishnan, R., Livny, M.: BIRCH: An efficient data clustering method for very large databases. ACM SIGMOD Record 25, 103–114 (1996)

Enabling Product Comparisons on Unstructured Information Using Ontology Matching

Maximilian Walther, Niels Jäckel, Daniel Schuster, and Alexander Schill

Abstract. Information extraction approaches are heavily used to gather product information on the Web, especially focusing on technical product specifications. If requesting different sources for retrieving such specifications, the outcome is of varying formats (different languages, units, etc.). The problem of how to bring such information sets into a unique, interchangeable format is not considered in many extraction systems. We develop a generic process for semantically integrating heterogeneous product specifications with the help of a product information ontology. The approach is based on a number of measures for detecting the right product attributes in the ontology to be matched with the extracted specifications and finally normalizing the specifications' values (e.g., concerning units). The feasibility of our approach is proven in a federated product search prototype called Fedseeko.

Keywords: information extraction, federated search, ontology matching, product information management.

1 Introduction

Today's World Wide Web offers a wide amount of product information with disparate structure and location which is not easy-to-handle for the average online consumer anymore. This led to a need for platforms offering effective product comparisons. In many cases, the product information maintained on such platforms is still acquired by hand. Figure 1 shows examples of technical specifications lists for two digital cameras *dc*1 and *dc*2 being represented

Maximilian Walther · Niels Jäckel · Daniel Schuster · Alexander Schill
Technische Universität Dresden, Faculty of Computer Science,
Institute of Systems Architecture, Helmholtzstr. 10, 01062 Dresden, Germany
e-mail: `maximilian.walther@tu-dresden.de`

E. Mugellini et al. (Eds.): Advances in Intelligent Web Mastering - 3, AISC 86, pp. 183–193.
springerlink.com

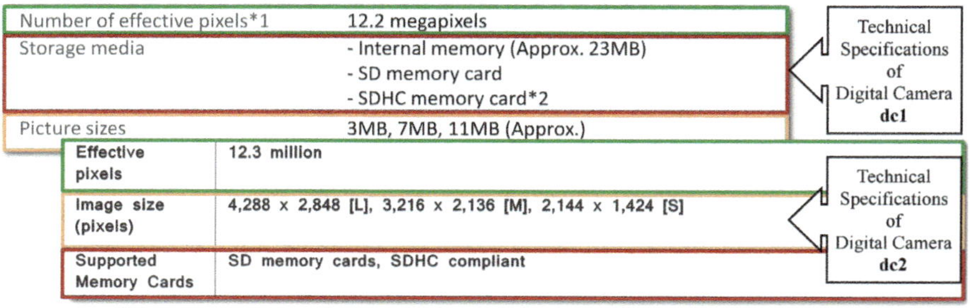

Fig. 1 Example for varying product specification formats of two digital cameras

in different producer-dependent manners. For being able to compare both products on a dedicated platform, an employee has to extract those specifications, match them with a corresponding product model and normalize included values.

Hence, many approaches for automating the product information collection on the Web have been developed which mostly focus on extracting product information from unstructured or semi-structured sources like the producer websites shown in the figure. The part of integrating extraction targets with a central knowledge model is not considered in many cases. However, this step is important since it makes products comparable. This applies especially to electronic products where technical product specifications allow effective comparisons and strongly affect a potential consumer in choosing a particular product.

We develop a generic approach for integrating technical product specifications with a given product information ontology. Our contributions consist of an appropriate process for extracting, classifying, matching and normalizing product specifications using a central knowledge model and a number of domain-specific measures required for the product specification matching process. In the following, the topic of ontology matching will be emphasized.

2 Ontology Matching

Although product information is generally not available in the form of an ontology online, each web source's product information has a distinct structure described by some internal schema. Thus, it is reasonable to assume that extracted specifications are implicitely modeled by an ontology, shifting our problem to the area of ontology matching. *Ontology Matching*[1] describes the process of finding correlations (*Ontology Alignment*) between entities of different ontologies that can be used for transforming one ontology into another (*Ontology Mapping*). Ontology matching is closely related to schema matching. The characteristic sequence for matching schemas consists of the actual *matching* step, an *aggregation* and a *selection* step. Newer systems[2] diversify this sequence for offering more flexibility.

State of the art matching systems usually apply several elementary matchers for finding an alignment. Considering their matching granularity, they can generally be divided into element-level and structure-level matchers. A more detailed matcher categorization is given in [3]. Depending on whether the system executes its matchers sequentially or in parallel, the overall matcher is called hybrid or composite, respectively. When talking about extracted product specifications, structural information is not available, thus reducing the set of elementary matchers to the element-level ones. For compensating this major drawback, as many characteristics of product specifications as possible have to be identified to be exploited by adequate element-level matchers. The fact that not only the specifications' schema (tbox), but also corresponding instances (abox) are extracted, is helpful in this case.

2.1 Element-Level Matchers

The most basic element-level matcher type is the one of string-based matchers. Such matchers only compare given strings, e.g., by calculating the Levenshtein distance. COMA[4], a composite schema matcher that allows the combination of different matching algorithms, includes four such matchers. Cupid[5] uses two different string-based matchers for detecting prefixes and suffixes in its first operation step. String-based matchers are used in nearly every matching system.

Language-based matchers use NLP techniques for identifying individual words or phrases or execute a morphological analysis on given strings. In S-Match[6] such matchers are used for detecting the meaning of given concepts.

The third type are matchers using linguistic resources such as WordNet, e.g., for retrieving synonyms of given schema strings. OWL Lite Aligner[7] (OLA) uses WordNet while systems like Naive Ontology Mapping[8] (NOM) and its successor Quick Ontology Mapping[9] (QOM) apply application-specific vocabularies. If the vocabulary used in a schema is not too particular for a special domain, such matchers can be quite powerful.

Constraint-based matchers have the ability to detect similar datatypes or multiplicities. E.g., such a matcher could figure out that a datatype day and workingday are quite similar. Similarity Flooding[10] is a hybrid schema matching system and uses constraint-based matchers while doing fix-point computations on the graph representations of its input schemas.

The last element-level matcher type being interesting for this work is the one of alignment reusing. Such matchers try to find alignments between the input schemas and other schemas. If for example both input schemas would already have been matched with a third schema and the resulting alignments are known to the matcher, a transitive mapping approach would support the matching process. COMA was the first system to offer alignment reuse.

Although the matchers presented here are taken from a survey[3] of schema-based matching approaches, they may as well be applied to instance data. Thus, each similarity measure presented later-on will reference one of these matcher types. In the following, systems directly working on instance data are described.

2.2 Instance Matching

Instance matching has been adopted in a set of different approaches. ASID[11] is a relational database schema mapper. It divides its matchers in strong and weak matchers while the weak matchers exploit available instance data that is to be cleaned in advance. Automatch[12] is solely based on instance data of different schemas. The instance data is compared with a unique internal schema using techniques of the machine learning domain. GLUE[13] is a system for semi-automatic taxonomy and ontology matching. The first of two basic learning strategies is the content learner that captures word counts of instance data.

These systems are designed quite generically for being able to match a lot of different schemas. Their approaches can be adapted to better fit our product specifications domain.

3 Semantic Integration of Product Specifications

The objective of our system is to create integrated sets of product specifications originally gathered on the Web for being able to compare products efficiently. The overall process is presented in Figure 2.

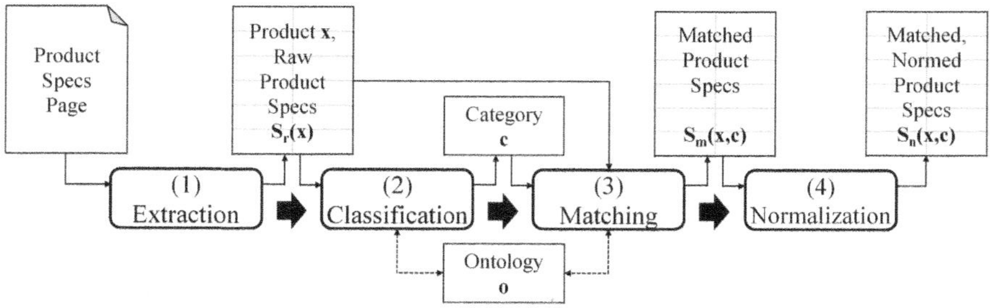

Fig. 2 Overall extraction and matching process

As can be seen, the first step consists in extracting the set of raw technical specifications $S_r(x)$ (e.g., "Number of effective pixels*1: 12.2 megapixels" for $dc1$ in Figure 1) of a product x (in this case, our $dc1$) from a dedicated web page (e.g., a page on $dc1$'s producer website). This step as well as the automated locating of such web pages has already been described in [14]. Since the product's category might be unknown, the second step consists in classifying the product to be located in a product category $c \in C$ (e.g., "Digital Cameras") from an ontology o. This step is essential since the matching algorithm needs to know which properties of o can potentially be matched with $S_r(x)$. The basic approach of how to classify products has already been described in [15]. We developed a number of refinements for this technique. However, since it is not the main focus of this paper, it will only be taken into account for the evaluation section. In the main step, the extracted specifications can be matched with c's set of abstract product specifications (e.g., "Image Resolution"). For disambiguation reasons, the abstract product specifications of o will be called the set of properties P in the following. The output of the matching step consists of a set of specifications $S_m(x,c)$ (e.g., "Image Resolution = 12.3 MP") being matched with and normalized by c. Finally, since each product specifications source uses its own style to represent specification values, a normalization of specification values is required.

Before examining the matching step in detail, the central concepts of our matching ontology are to be presented since they will be used throughout the whole process.

3.1 Domain Model

The ontology o represents the target schema for matching extracted product specifications $S_r(x)$ and was modeled in OWL. Only the tbox of o is of interest. Generally, an ontology tbox for the product domain would contain a taxonomy of product types, relations between those types and corresponding product attributes. However, since we introduced the concept of a *property* in our matching process, we created a tbox meta-model being located above the normal tbox model that allows the description of product property attributes, such as the property's structure, type, etc. The meta-model is presented in Figure 3.

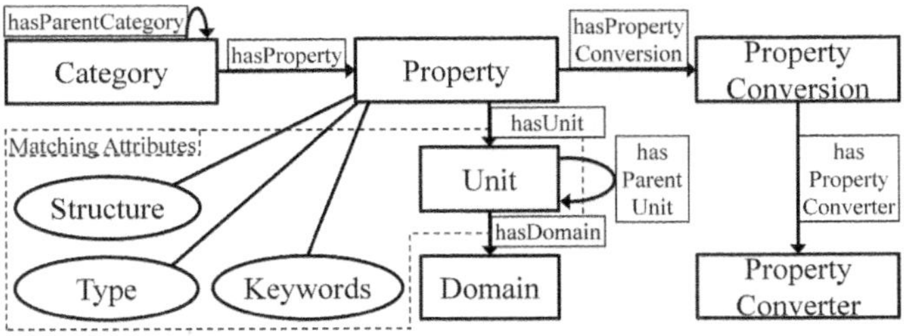

Fig. 3 Domain meta-model for the matching process

As shown by the figure, the meta-model mainly consists of the mentioned categories C, corresponding properties P, a set of units U and a description of each unit's domain. The lower tbox model contains categories such as

"Digital Cameras", properties such as "Image Resolution" or "Image Sizes", and units including "Metre", etc. The relations between those concepts can be seen in the figure as well. Additionally, each property has three different attributes, namely, a structure, a type and a collection of keywords. These attributes as well as the property's unit will be important for the matching process to be presented later on. An additional product conversion attribute is provided that describes how a property can be split up into basic properties or combined to a complex property using a property converter (a dedicated code snippet for executing the conversion process). These two classes are important for the normalization process.

The described meta-model as well as concrete categories and properties for those categories were modeled in OWL and build the tbox of o. For each category, up to 40 properties have been added. Figure 4 shows the specification set of $dc1$ after being matched with and normalized by our ontology.

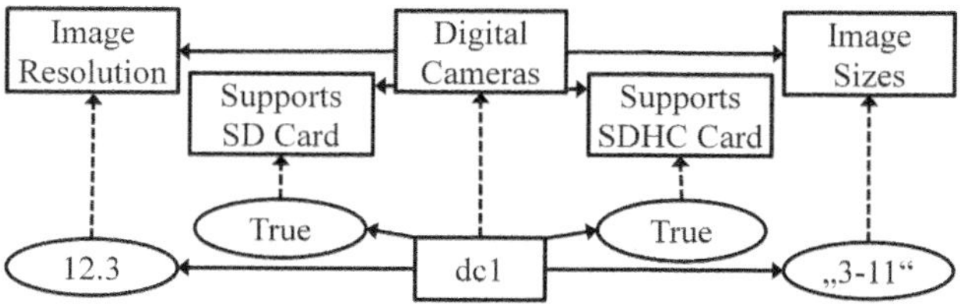

Fig. 4 Examples of matched product specifications for $dc1$ in Figure 1

3.2 Product Specification Normalization

If an alignment between an extracted specification s_r and a property p has been detected, the correct key of s_r is the one of p. However, if the extracted specification has a complex structure, such as a list, a vector, or a range, contained values are not easily comparable. Thus, the complex specification is split into several elementary and comparable specifications. Furthermore, the contained values are cleaned by removing HTML snippets, deleting non-relevant information (e.g., information in brackets), distilling numeric values or changing boolean values to "true" and "false", respectively. Furthermore, found units are changed to the current specification's standard unit followed by a recalculation of the associated numeric value (Figure 5).

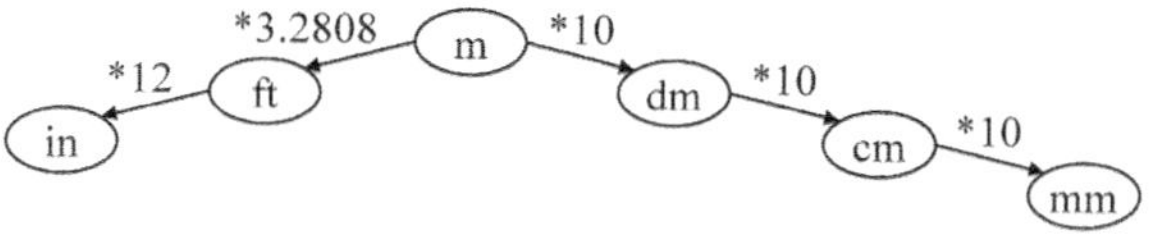

Fig. 5 Relations between length units

4 Similarity Measures

In the following, we develop a composite matcher that uses a set of element-level matchers specifically designed for the product domain. During the adoption of those matchers, various thresholds are employed to only select the most robust alignments. Finally, the resulting elementary alignments are aggregated.

We identified five characteristics of technical product specifications that may be used as indications for the matching process. One of them is the specification's key. The other characteristics focus on the specification's value and include the value's structure, the value's type, the value's unit and a collection of keywords potentially contained in the value. The following formula shows how to aggregate the different similarities to create a consolidated similarity value.

$$\Phi S_{spec}(s_r, p) = \Theta_{\tau_{spec}}\Big(\Phi S_{key}(s_r, p) + \Phi S_{struct}(s_r, p) + \Phi S_{type}(s_r, p) + \Phi S_{unit}(s_r, p) + \Phi S_{keyword}(s_r, p)\Big) \quad (1)$$

A function $\Theta_{\tau_{spec}}$ is used to define a threshold. Each of the different similarity functions is to be implemented in its own elementary matcher and will be defined below.

Key Similarity. The key similarity $\Phi S_{key}(s_r, p)$ is to be calculated through a string comparison. The most basic way to compare a specification key and a property key is to check whether both keys are identical. A value of 1 is used as similarity value in that case. If $key(s_r)$ is contained in $key(p)$ or vice-versa, the similarity is the ratio of both values multiplied with a weight between zero and one. Additionally, in the latter case, a threshold function sets the similarity to zero if the contained string is too short. If the keys are not identical and no key contains the other, the Levenshtein distance (normalized by the length of the longer key) is used. Again a weight and a threshold function decide about the overall value. Finally, if neither of both similarities is above zero, alignments from previous matching tasks are examined for detecting potential transitive mappings. The resulting matcher is therefore string-based and alignment reusing.

For the specifications of $dc1$ in Figure 1, the similarity of "Picture sizes" and the property name "Image Sizes" from our example tbox in Figure 4 could be detected with the Levenshtein function.

Structure Similarity. The first of four value similarity measures is the structure similarity. The structure of a product specification's value can assume four shapes, namely, range, vector, list and scalar. For being able to calculate the structural similarity of a product specification s_r and a property p from the ontology o, an extraction function $E : val(s_r), pat \mapsto val(s'_r), val(s'_r) \subseteq val(s_r)$ is needed that searches for patterns of all the mentioned shapes in a given specification value and extracts each part complying with such a pattern. Then, the length of the extracted pattern match is normalized by the complete value length.

The final similarity measure $\Phi S_{struct}(s_r, p)$ for the value's structure is the maximum of all four values multiplied with the structure's weight. If neither a range, nor a vector or list could be detected, the algorithm assumes to have found a scalar value. Since scalar values are quite meaningless, the similarity is set to zero in such cases. In Figure 1, a list could be detected when examining "Storage Media" or "Picture Sizes" from *dc*1. This would give a hint that "Storage Media" is the complex version of different properties including "Supports SD card". This complex property is not shown in the example tbox. A matcher implementing this similarity measure would belong to the contraint-based class.

Type Similarity. The type similarity $\Phi S_{type}(s_r, p)$ is the second value similarity measure and detects the datatype in a product specification's value. Thus, it is also part of a constraint-based matcher. We identified four types, namely, boolean, float, integer and string. The calculation is similar to the structure similarity measure. Again, by the use of an extraction function, datatype patterns are searched in the specification value. The highest extraction value is multiplied with the found datatype's weight. Since a string type is not that significant, its detection leads to a similarity of zero. Coming back to our example specifications in Figure 1, several datatypes could be detected such as a float value in "Number of effective pixels*1" or some integer values in "Picture sizes".

Unit Similarity. The unit similarity $\Phi S_{unit}(s_r, p)$ is also based on an extraction function that uses patterns for identifying units in given specification values. Since some units are more specific for a potential matching property, the result of the extraction function is multiplied with the found unit's specificity and a corresponding weight. A unit's specificity depends on how many properties in o use this property for their values. The more properties use a particular property, the less specific the unit is for each of these properties.

Product specifications value units may vary even if they belong to the same property (e.g., metres, centimetres, feet). Thus, a unit model has to be included in the product information ontology (Figure 5). The relations between units of the same domain allow a wider search for unit patterns. The formula's similarity measure is based on the best extraction result for all units related to the checked property while derived units are multiplied with a different weight. The unit similarity matcher belongs to the class of linguistic resources matchers.

Our running example includes "megapixels" in "Number of effective pixels*1" or "MB" in "Storage Media". These are hints for potentially matching properties.

Keyword Similarity. The last similarity we defined for matching product specifications with properties from an ontology is the keyword similarity

$\Phi S_{keyword}(s_r, p)$, also belonging to the class of linguistic resources matchers. For each keyword defined for a corresponding property p, a pattern matching function calculates a relative pattern matching value. The keyword similarity is the sum of all keyword matches. The $dc2$'s attribute "Supported Memory Cards" in Figure 1 includes keywords such as "SD" or "SDHC".

Having combined the different matchers for calculating the overall similarity value of a defined property p and an extracted specification s_r as shown above, an alignment $S_m(x, c)$ can be constructed taking the stable marriages problem into account. The normalization is the final step in our process chain.

5 Evaluation

We evaluated our matching process concerning seven different product categories from the electronics domain in a prototype called Fedseeko. First, a test set of products was used to determine the weights and thresholds of our algorithm empirically. Afterwards, 131 products equally distributed over all categories were gathered by a product crawler randomly selecting products from the Amazon portfolio and assembled manually to create a gold standard. Only product specifications being modeled as properties in our ontology o were taken into account. Then, our system tried to automatically execute the necessary matching tasks. Figure 6 shows the precision and recall values for product classification and property matching independently. The figure reveals that with our chosen thresholds the system did not always classify the given products (71.8%), but indeed for those 71.8% of the product set always chose the correct product category. The overall property matching gained quite similar values. Lower thresholds might have improved the recall for both the classification of products as well as the matching of extracted specifications at the expense of a worse precision.

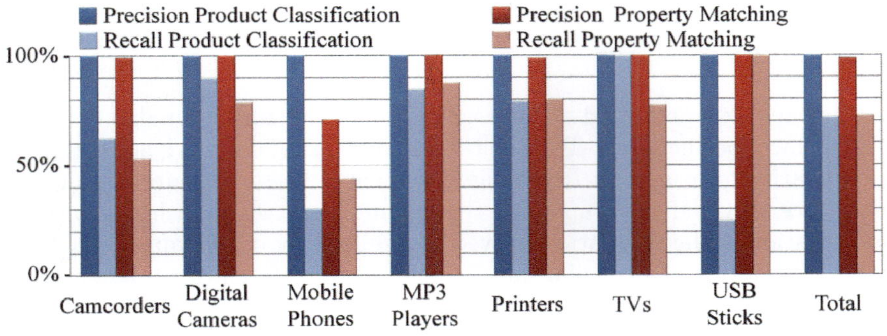

Fig. 6 Evaluation of precision and recall for classification and matching

Figure 7 displays the F-measure. Additionally, the quality of product specification normalization is displayed. The overall F-measure values account 83.6% for the classification and matching, 92% for the normalization and 64.3% for the overall process. This value falls out of alignment since the normalization step dependends on a successful execution of all previous steps.

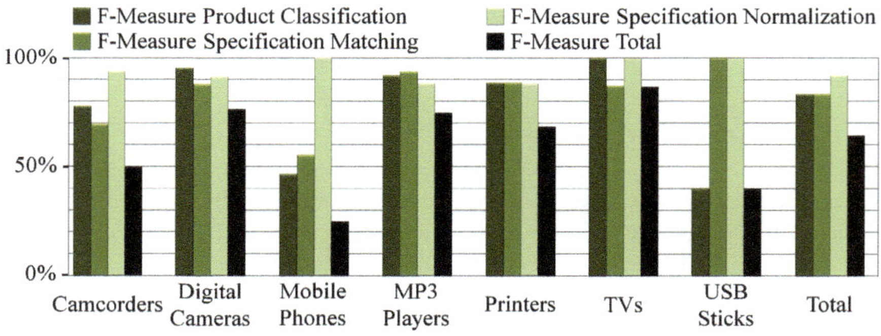

Fig. 7 Evaluation of F-measure values for the whole process

The developed approach was mainly employed for creating specification sets of electronic end-consumer products since for those products specifications can generally be found online. From the chosen categories, mobile phones and USB sticks showed the worst results. This is mainly due to the incomplete sets $P("MobilePhones")$ and $P("USBSticks")$ which impair the quality of our classification process. Missing keywords and units of existing properties have a negative impact on the specification matching itself. Thus, a deliberately modeled, fully-fledged ontology would already improve the evaluation results significantly.

6 Conclusions

In our paper, we presented an approach for matching and normalizing technical product specifications. The capital contribution of our work is a collection of similarity measures for detecting alignments. These measures are important since they help to compensate the missing hierarchy in technical product specifications. The evaluation proved the suitability of our approach. However, automating the calculation of employed weights and thresholds would certainly improve the flexibility of our system as well as overall results. A feasible approach would be the adoption of semi-supervised or unsupervised learning techniques. In addition, language-based matchers using NLP and a dynamic adaptation of the matching process could further enhance the evaluation results.

References

1. Euzenat, J., Shvaiko, P.: Ontology matching, 1st edn. Springer, Heidelberg (2007)
2. Peukert, E., Berthold, H., Rahm, E.: Rewrite techniques for performance optimization of schema matching processes. In: Proceedings of the 13th EDBT, pp. 453–464. ACM, New York (2010)
3. Shvaiko, P., Euzenat, J.: A survey of schema-based matching approaches. Journal on Data Semantics 4, 146–171 (2005)
4. Do, H.H., Rahm, E.: Coma - a system for flexible combination of schema matching approaches. In: Proceedings of the 28th VLDB, VLDB Endowment, pp. 610–621 (2002)
5. Madhavan, J., Bernstein, P.A., Rahm, E.: Generic schema matching with cupid. In: Proceedings of the 27th VLDB, pp. 49–58. Morgan Kaufmann Publishers Inc., San Francisco (2001)
6. Giunchiglia, F., Shvaiko, P., Yatskevich, M.: S-match: an algorithm and an implementation of semantic matching. In: Bussler, C.J., Davies, J., Fensel, D., Studer, R. (eds.) ESWS 2004. LNCS, vol. 3053, pp. 61–75. Springer, Heidelberg (2004)
7. Euzenat, J., Valtchev, P.: Similarity-based ontology alignment in owl-lite. In: Proceedings of the 16th ECAI, pp. 333–337. IOS Press, Amsterdam (2004)
8. Ehrig, M., Sure, Y.: Ontology mapping - an integrated approach. In: Bussler, C., Davies, J., Fensel, D., Studer, R. (eds.) ESWS 2004. LNCS, vol. 3053, pp. 76–91. Springer, Heidelberg (2004)
9. Ehrig, M., Staab, S.: QOM – quick ontology mapping. In: McIlraith, S.A., Plexousakis, D., van Harmelen, F. (eds.) ISWC 2004. LNCS, vol. 3298, pp. 683–697. Springer, Heidelberg (2004)
10. Melnik, S., Garcia-Molina, H., Rahm, E.: Similarity flooding: A versatile graph matching algorithm and its application to schema matching. In: Proceedings of the 18th ICDE, pp. 117–128 (2002)
11. Bozovic, N., Vassalos, V.: Two-phase schema matching in real world relational databases. In: Proceedings of the ICDE Workshops, pp. 290–296 (2008)
12. Berlin, J., Motro, A.: Database schema matching using machine learning with feature selection. In: Pidduck, A.B., Mylopoulos, J., Woo, C.C., Ozsu, M.T. (eds.) CAiSE 2002. LNCS, vol. 2348, pp. 452–466. Springer, Heidelberg (2002)
13. Doan, A., Madhavan, J., Dhamankar, R., Domingos, P., Halevy, A.: Learning to match ontologies on the semantic web. The VLDB Journal 12(4), 303–319 (2003)
14. Walther, M., Hähne, L., Schuster, D., Schill, A.: Locating and extracting product specifications from producer websites. In: Proceedings of the 12th ICEIS, INSTICC (2010)
15. Walther, M., Schuster, D., Juchheim, T., Schill, A.: Category-based ranking of federated product offers. In: Proceedings of the 8th WWW/Internet. IADIS Press (2009)

Analyzing Sentiment in a Large Set of Web Data While Accounting for Negation

Bas Heerschop, Paul van Iterson, Alexander Hogenboom, Flavius Frasincar, and Uzay Kaymak

Abstract. As virtual utterances of opinions or sentiment are becoming increasingly abundant on the Web, automated ways of analyzing sentiment in such data are becoming more and more urgent. In this paper, we provide a classification scheme for existing approaches to document sentiment analysis. As the role of negations in sentiment analysis has been explored only to a limited extent, we additionally investigate the impact of taking into account negation when analyzing sentiment. To this end, we utilize a basic sentiment analysis framework – consisting of a wordbank creation part and a document scoring part – taking into account negation. Our experimental results show that by accounting for negation, precision on human ratings increases with 1.17%. On a subset of selected documents containing negated words, precision increases with 2.23%.

Keywords: Sentiment analysis, negation, wordbank creation, document scoring.

1 Introduction

With the advent of the Web, traces of human activity and communication have become ubiquitous, partly in the form of written text. In recent years, virtual utterances of opinions or sentiment have become increasingly abundant through messages on Twitter, on-line customer reviews, etcetera. The information contained in this ever-growing data source of the Web is invaluable to key decision makers, e.g., those making decisions related to reputation management or marketing. An understanding of what is going on in their particular markets is crucial for decision makers, yet the analysis of sentiment in an overwhelming amount of data is far from trivial.

Bas Heerschop · Paul van Iterson · Alexander Hogenboom · Flavius Frasincar · Uzay Kaymak
Erasmus University Rotterdam, PO Box 1738, NL-3000 DR, Rotterdam, The Netherlands
e-mail: {basheerschop,paulvaniterson}@gmail.com,
{hogenboom,frasincar,kaymak}@ese.eur.nl

E. Mugellini et al. (Eds.): Advances in Intelligent Web Mastering - 3, AISC 86, pp. 195–205.
springerlink.com

Sentiment analysis refers to a broad area of natural language processing, computational linguistics and text mining. In general it aims to determine the attitude, evaluation, or emotions of the author with respect to the subject of the text. The basis of sentiment analysis is determining the positive-negative polarity of a text. The research area of sentiment analysis is relatively new, with many aspects being currently explored. The most promising areas of focus are word sentiment scoring (i.e., learning the sentiment scores of single words), subject/aspect relevance filtering (i.e., determining the subject and/or aspect a sentiment carrying word is relevant to), subjectivity analysis (i.e., determining whether a sentence is subjective or objective), and sentiment amplification and negation (i.e., modifying sentiment strength on amplifying words and reversing sentiment scores on negated words).

Some researchers have already suggested to account for negation when analyzing sentiment in texts. Yet so far, the impact of taking into account negation when analyzing sentiment has not been demonstrated. Therefore, we present our first steps towards insight in the impact of negation on sentiment analysis. The remainder of this paper is organized as follows. First, we classify existing sentiment analysis approaches and assess the extent to which they account for negation in Sect. 2. Then, we describe and utilize our framework for assessing the impact of negation in sentiment analysis in Sect. 3. We conclude in Sect. 4.

2 Sentiment Analysis

In recent years, several approaches to sentiment analysis (i.e., classification) of documents have been proposed. Most approaches essentially adhere to more or less similar frameworks. One of such frameworks is the basic framework proposed by Liu [11, 12], consisting of an algorithm for creating a wordbank (i.e., a list of words and their associated sentiment) from a training corpus, along with a document-level scoring function. Ceserano et al. [3] propose a similar framework, which has been used by other researchers as well [1, 2]: OASYS. OASYS provides two word scoring algorithms based on supervised learning and three sentence-level document scoring algorithms with topic relevance filtering. Despite adhering to similar frameworks, document sentiment analysis approaches have several characteristic features distinguishing them from one another. We consider the following features.

Wordbank (WB). Most approaches rely on a wordbank, typically containing per-word sentiment scores. Creation methods include supervised learning on a set of manually rated documents, learning through related word expansion (expanding a small, manually created set of words by exploiting word relationships such as synonyms, antonyms, and hypernyms), completely manual creation, or a combination of these methods. The target of the wordbank (e.g., general or domain-specific) may also differ amongst approaches, as well as the differentiation between part-of-speech variations of a word.

Sentiment scoring level (SSL). One could consider sentiment analysis to be performed at document level, sentence level, or window level.

Topic relevance filtering (TRF). Taking into account the way in which sentiment carrying words are tied to their subject results in allegedly irrelevant phrases being filtered out of further processing. More advanced methods look at the surrounding words of a sentiment carrying word, the subject of sentences that contain sentiment, or specific features of a subject.

Subjectivity filtering (SF). Subjective sentences carry sentiment, whereas objective sentences only carry factual information. Ignoring objective sentences is crucial for some sentence-level and window-level algorithms with aggregation functions averaging sentiment expressed in the subparts. Including objective sentences here would decrease the impact of subjective sentences.

Part-of-speech tagging (POS). Annotating words with their corresponding parts-of-speech (POS) – e.g., noun, verb, adjective, subject, or object – can help algorithms making better decisions. For example, "I like A", where "like" is a verb carrying high positive sentiment, is very different from "A is like B" where "like" is an adverb carrying no sentiment. POS tagging can also be used to identify the subject of a sentence to which a sentiment carrying adjective or verb applies. Additionally, words that cannot carry sentiment can be filtered; only adjectives, adverbs, verbs and nouns carry sentiment [2, 12, 15].

Negation (NEG). Linguistic negation is the process that turns an affirmative statement ("I like A") into the opposite denial ("I do not like A"). In general, negation is done by the inclusion of a negation keyword (e.g., "not" or "never"), but negation can also be achieved using clauses like "but" ("Feature A is excellent, but feature B ..."). An important aspect in negation is the identification of the sentiment carrying word the negation applies to [12].

Amplification (AMP). The process of increasing or decreasing the sentiment score of a word, when it is combined with an amplification word, is typically done by multiplying the sentiment score of the word by the amplification score of the amplification word. For example, the positive sentiment score of "beautiful" would be increased by multiplying it with the amplification score of "very" in "very beautiful".

Comparison (COMP). An author's sentiment on a topic, relative to his sentiment on another topic can be determined by means of comparison (e.g., "A is better than B"). In sentiment analysis, the absolute sentiment of the author on a topic is typically extracted ("A is good"). Relative comparative sentiment analysis can only be converted to absolute values if an absolute sentiment analysis can be done. For example, if the sentiment score on A can be determined, and if we know that "A is better than B", we can deduce that the sentiment score on B must be lower than the sentiment score on A, with an amount depending on the strength of the comparison.

Syntactical variants (SYN). Reducing the variability in the forms of words as much as possible can increase the accuracy of word counts. Words can, for example, have alternative spellings or spelling errors. Also verbs, adjectives, and adverbs can be transformed grammatically. Stemming and lemmatizing are techniques to bring back transformed words to their base form (e.g., bring "loved" in "I loved it" back to its stem "love").

Based on these features, the state-of-the-art in sentiment analysis can be characterized. Table 1 presents an overview of several recent approaches. All three OASYS document scoring algorithms introduced by Ceserano et al. [3] use a wordbank created through supervised learning. All methods do TRF; the Topic-Focussed (TF) algorithm only handles sentences that contain a reference to the topic, the Distance-Weighted Topic-Focussed algorithm (DTWF) gives more weight to sentiment near topic keywords (and is hence a window-level approach), and the Template-Based algorithm (TB) only handles sentences that match a certain template (e.g.., sentence structure or keywords). Besides POS tagging, the OASYS algorithms do not support any other features.

Lerman et al. [10] propose three sentence-level sentiment summarization algorithms: Sentiment Match (SM), Sentiment Match and Aspect Coverage (SMAC), and Sentiment Aspect Match (SAM). The algorithms compute a textual summary of the input document, where sentences are selected to maximize total sentiment in the summary. The algorithms use a wordbank, which is created by related word expansion from a manually annotated base collection using WordNet [6]. By selecting sentences with the highest absolute sentiment score, objective sentences are filtered out. All algorithms filter for topic relevance, where SMAC and SAM use a more advanced, feature-based approach.

The Adverb-Adjective Combinations (AACs) proposed by Benamara et al. [2] use a linguistic analysis of adverbs of degree, which modify adjectives (e.g., "very beautiful"). The algorithms – Variable Scoring (VS), Adjective Priority Scoring (APS), and Adverb First Scoring (AFS) – vary in how they weight the adverb amplification scores. The AAC framework builds on the OASYS framework. The differences between the original OASYS implementation and the AAC implementation are that the AAC implementation supports negation and amplification, and requires a second wordbank containing adverb amplification scores.

Ding et al. [5] propose a holistic lexicon-based sentence-level sentiment analysis approach. Their Opinion Observer (OO) handles context-dependent opinion words

Table 1 Classification of algorithms

Algorithm	WB	SSL	TRF	SF	POS	NEG	AMP	COM	SYN
TF [3]	yes	sentence	yes	no	yes	no	no	no	no
DWTF [3]	yes	window	yes	no	yes	no	no	no	no
TB [3]	yes	sentence	yes	no	yes	no	no	no	no
SM [10]	yes	sentence	yes	yes	no	no	no	no	no
SMAC [10]	yes	sentence	yes	yes	no	no	no	no	no
SAM [10]	yes	sentence	yes	yes	no	no	no	no	no
VS [2]	yes	sentence	yes	no	yes	yes	no	no	no
APS [2]	yes	sentence	yes	no	yes	yes	no	no	no
AFS [2]	yes	sentence	yes	no	yes	yes	no	no	no
OO [5]	yes	sentence	yes	yes	yes	yes	no	no	no
CSR [9]	no	document	yes	no	yes	no	no	yes	no
EVAL [11]	yes	document	no	no	no	no	no	no	no

and deals with many special words, phrases and language constructs which impact opinions through their linguistic patterns. OO uses a wordbank that is created using related work expansion (via WordNet) on a small set of manually annotated words. Subjectivity filtering is done by ignoring sentences that do not contain sentiment-carrying words. Linguistic negation is recognized through negation words, which include traditional keywords (e.g., "not") and pattern-based negations such as "stop" + verb + "ing" (e.g., "stop liking").

Rather than using a wordbank with absolute word sentiment scores, the Class Sequential Rules (CSR) approach proposed by Jindal and Liu [9] uses sequential pattern mining to identify sub-sequences of text that occur more often than a minimum support threshold. The patterns used as features consist of POS tags and one or more comparative key phrases. The sentiment orientation of the key phrases determines the orientation of a sequential pattern. For example, the phrase "Intel is better than AMD" yields the comparative pattern {{proper noun} {third person singular present tense verb} {"better", comparative adjective} {subordinating preposition or conjunction} {proper noun}}.

The Evaluate document algorithm (EVAL) proposed by Liu [11] implements a very basic sentiment analysis framework. It works on the document level and sums up all the individual word sentiment scores, stored in a wordbank, to compute the document score. It does not support any of our classification properties.

Most approaches agree that adjectives and adverbs carry the most sentiment. The role of negations has been explored only to a limited extent. Therefore, we propose to shed some light onto the impact of accounting for negation in sentiment analysis.

3 Sentiment Negation

In order to assess the impact of sentiment negation, we propose a very simple sentiment analysis framework, similar to Liu [11]. This framework consists of wordbank creation and subsequent lexicon-based document scoring. Both parts have optional support for sentiment negation. We classify a document as either positive (1), neutral (0), or negative (-1). The score range of individual words is [-1, 1]. Our framework focuses on adjectives, as adjectives are the best indicators of sentiment [2, 12, 15].

3.1 Framework

The first part of our framework facilitates wordbank creation, involving scoring sentiment of individual words (adjectives) w in a training corpus D_{train}. Our word scoring function is based on a pseudo-expected value function [3]. The sentiment score of any adjective w, $\text{score}(w)$, is based on its total relative influence on the sentiment over all documents $d \in D_w$, where $D_w \subseteq D$, with each document containing w:

$$\text{score}(w) = \frac{\sum_{d \in D_w} \text{score}(d) \times \inf(w, d, neg)}{|D_w|}, \tag{1}$$

where $\text{score}(d)$ is an individual document d's manually assigned score, $|D_w|$ is the number of documents in D_w, and $\inf(w,d,neg)$ is the relative influence of an adjective w in document d, with a Boolean neg indicating whether to account for negation or not. This influence is calculated as the count $\text{freq}(w,d,neg)$ of w in d in relation to the total frequency $\sum_{w' \in d} \text{freq}(w',d,neg)$ of all sentiment carrying words w' in d:

$$\inf(w,d,neg) = \frac{\text{freq}(w,d,neg)}{\sum_{w' \in d} \text{freq}(w',d,neg)}. \quad (2)$$

In order to support negation in our framework, we use a variation of Hu and Liu's method [7] of negation. Yet, even though optimizing the scope of influence of negation words [8, 13, 14] or exploitation of compositional semantics in sentiment-bearing expressions [4] has its merits, we first focus on a one-word scope for negation words in an attempt to tease out the effects of accounting for even the simplest forms of negation, as opposed to not accounting for negation at all. We only handle negation words that precede a sentiment word, as larger distances might cause noise in our results due to erroneously negated words. Support for negation is considered in the frequency computations by subtracting the number of negated occurrences of word w in d from the number of non-negated occurrences of w in d.

In the second part of our framework, the score $\text{eval}(d)$ of a document d containing n adjectives $\{w_1, w_2, \ldots, w_n\}$ is simply computed as the sum of the scores of the individual adjectives (the same adjective can appear multiple times), as determined using (1) and (2). In case negation is accounted for, we propose to use a document scoring function based in the scoring function presented by Liu [11]:

$$\text{eval}(d) = \sum_{w_i \in d} (-1)^{\text{negated}(w_i,d)} \times \text{score}(w_i), \quad (3)$$

where $\text{negated}(w_i,d)$ is a Boolean indicating whether the ith adjective in w is negated in d (1) or not (0). Using (3), the classification $\text{class}(d)$ of a document d can finally be determined as follows:

$$\text{class}(d) = \begin{cases} 1 & \text{if } \text{eval}(d) > 0.002, \\ 0 & \text{if } -0.021 \leq \text{eval}(d) \leq 0.002, \\ -1 & \text{if } \text{eval}(d) < -0.021. \end{cases} \quad (4)$$

In order to determine the optimal thresholds for (4), we have experimented with different values for the upper and lower threshold. For the upper threshold we have experimented with values between 0.001 and 0.5 with a step-size of 0.005. For the lower threshold, we have experimented with values between -0.001 and -0.5 with a step of 0.005. The ranges between which we tested were determined by manual analysis, in which we found results to decrease rapidly outside interval $[-0.5, 0.5]$.

3.2 Implementation

We have implemented our framework in C#, combined with a Microsoft SQL Server database. We have used a corpus of 13,628 human-rated Dutch documents on 40 different topics. Sentiment in these documents is classified as positive, negative, or neutral. In order to be able to asses the impact of negation, we have implemented two versions of our framework. The first version has no support for negation, whereas the second version supports negation both in the wordbank creation and in the document scoring part. Our framework only handles adjectives for sentiment analysis and uses a commercial part-of-speech tagger (based on OpenNLP and trained on Dutch corpora) to identify adjectives in the corpus.

We have used 60% of our documents for training and 40% for testing. The training set was used to create wordbanks and to determine the best threshold level for document classification. Our software first uses Algorithm 1 to retrieve all adjectives from the training corpus, where multiple occurrences of an adjective are not allowed. The list of adjectives thus extracted is subsequently used for creating a wordbank, hereby following Algorithm 2, which scores all adjectives occurring more than once in the training set with word scoring function (1). A Boolean variable is used to turn the support for negation on or off. Algorithm 3, in which support for negation can also be toggled, is subsequently used to score documents in accordance with document scoring functions (3) and (4).

3.3 Evaluation

In order to evaluate the human judgements, we took a random sample of 224 documents and rated these for sentiment. We observed 56% strong agreement and 99% weak agreement between our judgement and the human annotations, where strong agreement means an exact match and weak agreement means that one rating is positive or negative, whereas the other is neutral. Interestingly enough, in 17% of the cases where our ratings do not strongly agree, human raters appear to tie sentiment to the consequences of facts, which we call "factual sentiment". For example, the objective and hence neutral statement "Stock prices for our company went down 2%

Algorithm 1: Creating a list of adjectives

input : A training corpus D_{train}
output: A list *wordList* of all adjectives in D_{train}

$wordList = \emptyset$;
foreach d *in* D_{train} **do**
    $adjList$ = `getAdj(`d`); // Retrieve all adjectives in` d
    foreach adj *in* $adjList$ **do**
        if $adj \notin wordList$ **then** $wordList = \{wordList, adj\}$;
    end
end

Algorithm 2: Creating a wordbank

```
input : A training corpus D_train, a list wordList of all adjectives in D_train, and a
        Boolean neg indicating whether to account for negation
output: A list wordbank containing all adjectives in D_train with their scores
wordbank = ∅;
foreach w in wordList do
    |D_w| = 0; // Number of documents containing w
    sumWContr = 0; // Sum of all contributions of w in D_w
    foreach d in D_train do
        // Retrieve frequency of w in d, minus negated
        // occurrences, if neg
        freqWD = freq(w, d, neg); // Number of occurrences of w in d
        scoreD = getScore(d); // Human annotators' score for d
        if freqWD > 0 then
            |D_w| = |D_w| + 1;
            if scoreD ≠ 0 then
                sumAllWD = 0; // Count of all words of wordList in d
                foreach w' in wordList do
                    sumAllWD = sumAllWD + freq(w', d, neg);
                end
                infWD = freqWD / sumAllWD; // Influence of w in d
                sumWContr = sumWContr + (infWD × scoreD);
            end
        end
    end
    if |D_w| > 1 then
        scoreW = sumWContr / |D_w|;
        wordbank = {wordbank, {w, scoreW}};
    end
end
```

today" is judged as carrying (negative) sentiment. Another explanation for the discrepancies between ratings are interpretation differences. It is for instance difficult for humans to pick up on subtle cases of sentiment, which can be expressed in irony and tone. The interpretation of such subtle uses of sentiment can differ from person to person. The two cases of strong disagreement are due to misinterpretation of the text.

Additionally, we have evaluated the performance of our framework against human ratings in two set-ups: one with support for negation and one without support for negation. Precision improves with 1.17% from 70.41% without taking into account negation to 71.23% when accounting for negation. This observed improvement is even more evident when our framework is applied to a subset of the corpus, each document of which contains negated words (not necessarily adjectives). On this subset, precision increases with 2.23% from 69.44% without accounting for

Algorithm 3: Scoring a document

input : A list *wordbank* containing all adjectives in the training corpus with their scores, an upper threshold *utreshold* indicating the score above which a document is considered to be positive, a lower document score threshold *lthreshold* below which a document is considered to be negative, a Boolean *neg* indicating whether to account for negation, and a document *d*

output: A document score *result*

```
result = 0; // Final score for document d, initialized as neutral
docScore = 0; // Score for document d
adjList = getAdj(d); // Retrieve all adjectives in d
foreach adj in adjList do
    if adj ∈ wordbank then
        if neg then docscore = docscore + (−1)^isNegated(adj) × score(adj);
        else docscore = docscore + score(adj);
    end
end
if docscore > uthreshold then result = 1;
else if docscore < lthreshold then result = −1;
```

negation to 70.98% when taking into account negation. These results are notable given that only 0.85% of the sentences in the original corpus contain negations.

4 Conclusions and Future Work

The main contribution of this paper is two-fold. First of all, we have provided a characterization of current approaches to sentiment analysis, based on their wordbank type, sentiment scoring level, topic relevance filtering, subjectivity filtering, part-of-speech tagging, negation, amplification, comparison, and type variations. In this analysis, it has become apparent that the role of negations in sentiment analysis has been explored only to a limited extent.

The second contribution of this paper lies in our reported endeavors of shedding some light onto the impact of accounting for negation in sentiment analysis. Firstly, we have found that human raters tend to rate the consequences of factual information as carrying sentiment; an observation that may be taken into account in future work. Furthermore, our experiments with a basic sentiment analysis framework show that a relatively straightforward approach to accounting for negation already helps to increase precision with 1.17%. On a subset of selected documents containing negated words, precision increases with 2.23%. This is a notable result if we consider the fact that negation is sparsely used in our data set.

Nevertheless, it appears to be worthwhile to investigate the effects of optimizing the scope of influence of negation words in order to obtain more detailed insights in the impact of negation in sentiment analysis. We would also like to experiment with other types of words in our wordbank (e.g., adverbs, possibly combined with adjectives). Finally, we plan on taking into account degrees of negation. For instance, "not bad" is not necessarily "good", yet more likely slightly less positive than "good". All in all, a rather simple way of accounting for negation in sentiment analysis already helps to improve performance, yet we envisage that future work in the suggested directions could advance the state-of-the-art in sentiment analysis.

Acknowledgements. We would like to thank Teezir (`http://www.teezir.com`) for their technical support, fruitful discussions, and for supplying us with data for this research.

References

1. Bautin, M., Vijayarenu, L., Skiena, S.: International Sentiment Analysis for News and Blogs. In: 2nd International Conference on Weblogs and Social Media (ICWSM 2008), pp. 19–26. AAAI Press, Menlo Park (2008)
2. Benamara, F., Cesarano, C., Reforgatio, D.: Sentiment Analysis: Adjectives and Adverbs are better than Adjectives Alone. In: 1st International Conference on Weblogs and Social Media (ICWSM 2007), pp. 203–206. AAAI Press, Menlo Park (2007)
3. Cesarano, C., Dorr, B., Picariello, A., Reforgiato, D., Sagoff, A., Subrahmanian, V.: OASYS: An Opinion Analysis System. In: AAAI Spring Symposium on Computational Approaches to Analyzing Weblogs (CAAW 2006), pp. 21–26. AAAI Press, Menlo Park (2006)
4. Choi, Y., Cardie, C.: Learning with Compositional Semantics as Structural Inference for Subsentential Sentiment Analysis. In: 13th Conference on Empirical Methods in Natural Language Processing (EMNLP 2008), pp. 793–801. ACL (2008)
5. Ding, X., Lu, B., Yu, P.S.: A Holistic Lexicon-Based Approach to Opinion Mining. In: 1st ACM International Conference on Web Search and Web Data Mining (WSDM 2008), pp. 231–240. ACM, New York (2008)
6. Fellbaum, C.D.: WordNet: An Electronic Lexical Database. MIT Press, Cambridge (1998)
7. Hu, M., Liu, B.: Mining and Summarizing Customer Reviews. In: 10th ACM SIGKDD International Conference on Knowledge Discovery and Data Mining (KDD 2004), pp. 168–177. ACM, New York (2004)
8. Jia, L., Yu, C., Meng, W.: The Effect of Negation on Sentiment Analysis and Retrieval Effectiveness. In: 18th ACM Conference on Information and Knowledge Management (CIKM 2009), pp. 1827–1830. ACM, New York (2009)
9. Jindal, N., Liu, B.: Identifying Comparative Sentences in Text Documents. In: 29th Annual International ACM SIGIR Conference on Research and Development in Information Retrieval (SIGIR 2006), pp. 244–251. ACM, New York (2006)
10. Lerman, K., Blair-Goldensohn, S., McDonald, R.: Sentiment Summarization: Evaluating and Learning User Preferences. In: 12th Conference of the European Chapter of the ACL (EACL 2009), pp. 514–522. ACL (2009)
11. Liu, B.: Web Data Mining. Springer, Heidelberg (2007)
12. Liu, B.: Handbook of Natural Language Processing. In: Sentiment Analysis and Subjectivity, 2nd edn., pp. 627–667. CRC Press, Boca Raton (2010)

13. Morante, R., Liekens, A., Daelemans, W.: Learning the Scope of Negation in Biomedical Texts. In: 13th Conference on Empirical Methods in Natural Language Processing (EMNLP 2008), pp. 715–724. ACL (2008)
14. Morante, R., Daelemans, W.: A Metalearning Approach to Processing the Scope of Negation. In: Thirteenth Conference on Computational Natural Language Learning (CoNLL 2009), pp. 21–29. ACL (2009)
15. Wiebe, J.M.: Learning Subjective Adjectives from Corpora. In: 17th National Conference on Artificial Intelligence and Twelfth Conference on Innovative Applications of Artificial Intelligence (AAAI 2000), pp. 735–740. AAAI Press, Menlo Park (2000)

A Quality Assurance Framework for Ontology Construction and Refinement

Mamoru Ohta, Kouji Kozaki, and Riichiro Mizoguchi

Abstract. The quality of an ontology is an important factor that determines its utility. In order to assure its quality, in addition to form-based evaluation as to whether the ontology being constructed is written properly in terms of its form (syntax), content-based evaluation as to whether the ontology properly represents the target domain, whether the ontology actually serves for problem solving, etc. is also necessary. In this study, we investigate a framework for quality assurance of ontologies in Hozo, which is an environment for building/using ontologies that are being developed by the authors. As form-based evaluation, Hozo provides various assistance functions for properly editing an ontology in compliance with the rules. As content-based evaluation, Hozo introduce a method for supporting ontology evaluation thorough conceptual maps which are generated according to the user's viewpoint.

Keywords: building ontologies, ontology evaluation, development support system.

1 Introduction

With the recent trend toward the increasing use of information in technical domains, a strong demand is arising for systematization of knowledge in various technical domains. Ontology engineering is an approach of interest for systematization of knowledge, and ontologies are being constructed in various domains such as medical science, bioinformatics, nano-technology, education, environmental engineering and so on. With this background, there is a demand for development of methodologies and tools for assisting the construction of good ontologies by experts in individual domains, as well as ontology experts.

The quality of a constructed ontology is an important factor that determines its utility. In order to assure the quality of an ontology, it is necessary to evaluate

Mamoru Ohta · Kouji Kozaki · Riichiro Mizoguchi
The Institute of Scientific and Industrial Research (ISIR),
Osaka University, 8-1 Mihogaoka, Ibaraki, Osaka, 567-0047, Japan
e-mail: {ohta,kozaki,miz}@ei.sanken.osaka-u.ac.jp

E. Mugellini et al. (Eds.): Advances in Intelligent Web Mastering - 3, AISC 86, pp. 207–216.
springerlink.com

whether the ontology is written properly and to reflect the result in the construction task. Generally, ontology evaluation is roughly classified into two kinds: form-based (syntax) evaluation and content-based (semantic) evaluation. In the form-based evaluation, generally, formal approaches are employed; for example, in the case of the Web Ontology Language (OWL), contradictions are detected through reasoning processing based on description logic (DL). In the content-based evaluation, on the other hand, manual approaches are employed, such as demonstrating validity by applying a model based on the ontology to actual problem solving in the target domain or by creating an organization of domain experts for content-based evaluation[1].

This paper discusses a framework that is being developed for quality assurance of ontologies in Hozo[2], an environment for building/using ontologies that is being developed by the authors. Sect. 2 discusses a framework of editing assistance and consis-tency verification, which is introduced in Hozo for the purpose of form-based evaluation. Sect. 3 discusses a refinement assisting method based on a concept map, which is introduced in Hozo for the purpose of content-based evaluation, with several examples of its use in practice. Sect. 4 discusses comparison with related work. Lastly, Sect. 5 summarizes the achievements of this research and concludes this paper with further issues to be addressed.

2 A Framework for Form-Based Ontology Evaluation

As a quality assurance effort in the construction phase, it is necessary to verify whether the ontology that has been constructed is written properly in terms of its form (syntax). At this time, as well as whether the individual concept definitions are written properly in compliance with the description form, it is necessary to verify whether the ontology as a whole does not include inconsistencies among the concept definitions. A construction tool requires assisting functions mainly for properly editing an ontology in compliance with the rules. In order to perform form-based evaluation of an ontology by using a computer, there are two approaches: one is to perform checking in advance during editing and present errors, and the other is to verify consistency after editing and correct errors. In the framework adopted by Hozo, it is possible to use these two approaches selectively case by case; i.e., emphasis is placed on editing assistance during editing.

Thus, in this research, first, items that are to be verified in relation to the ontology rules were enumerated (32 items in total). As a function for assisting ontology editing, the items were classified into items that should at least be satisfied during editing, with which preferably errors should be presented as soon as detected (25 items), and items with which the consistency of the ontology as a whole is to be verified after editing and errors are to be presented after editing (27 items).

Hozo employs goal-oriented reasoners to assure fast consistency-check at the cost of completeness. Therefore, Hozo has 32 native reasoners corresponding to each item. We believe each of the reasoners is straightforward and needs no detail explanation. Apparently, all the reasoners are fired when consistency-check is done after editing, while some of them are fired in the course of editing process.

2.1 Ontology Editing Assistance

Hozo provides various assistance functions that help the user who constructs an ontology to properly edit the ontology in compliance with the rules. For example, when assisting slot specialization, in order to write the definition properly in compliance with the rules, during editing of a slot, a list of candidates of slots that can be inherited from the upper concept is presented so that the user can select and edit one of the candidates. On the other hand, in the case of some particular items, such as reference to an undefined concept, instead of requiring proper description, for example, an alert is presented in a different color, so to make the disturbance of the users thinking minimum. Furthermore, Hozo has an assisting function that prevents loss of consistency among concepts during modification of the description, such as modification of a concept name or modification of an inheritance relationship.

Hozo assures the quality of an ontology by using assistance functions for maintaining consistency so that the ontology doesn't include contradictions when existing concepts or slots are modified, as well as when new concepts are defined.

2.2 Consistency Verification after Ontology Editing

The editing assistance described above serves to prevent errors to some extent; however, there are errors that cannot be prevented since it is not possible to check consistency before finishing editing of a single concept. For example, while modification and deletion of concept definitions are repeated, in some cases, the relationship among the concepts is modified, resulting in loss of consistency of the ontology as a whole. Furthermore, there are cases where errors for which alerts were issued but that were disregarded during editing remain without being corrected. In order to overcome these problems, a scheme that ensures the consistency of the ontology as a whole after editing is necessary. Thus, the authors implemented functions for verifying the consistency of the ontology as a whole after editing and for presenting the results regarding items that are to be verified but were not covered by the assistance during editing. The verification results are presented as a list and the user can collect them by using a special wizard which assists correction of verification errors.

These consistency verification functions were developed by using ontology processing APIs HozoCore and Reasoner, which are software modules developed for computer processing of a Hozo-specific theoretical framework.

2.3 Evaluation of Ontology Editing Assistance

As an experiment for evaluating improvement in ontology quality in terms of formal aspects rather than content validity, we conducted a comparative experiment as to whether any improvement in the quality of the ontology construction work was achieved between before and after the implementation of the editing assistance functions described in Sect. 2.1. The subjects were graduate students with no advance knowledge of ontology. The procedure of the experiment was as follows. First, an explanation of ontology and a lecture of how to use Hozo were given in advance. Then, regarding a certain theme (vehicle ontology), the subjects were formed into groups each consisting of three subjects in such a manner that personal variation of knowledge was minimized, and each group constructed an ontology of vehicle using Hozo. In order to verify quality improvement, in the evaluation experiment, the same task was undertaken before and after the improvement with the tool to different subjects in different years. As the results of the experiment, the number of concepts, the number of slots (using inheritance or not), the number of relational links, and the number of errors in the constructed ontology were counted. Regarding the number of errors, the ontology consistency verification function described in Sect. 2.2 was used. In general, a quantity of concepts and slots in ontology building has little to do with quality of the ontology. But in this experiment, because subjects are true beginners who had never built any ontology, the primary issue for them was to acquire basic skills as to how to define concepts in consideration of attribute inheritance, and hence quality of ontology is the secondary issue. Therefore, if they could define more concepts and slots which are regarded as rather meaningful using property inheritance from their super concepts, it could be understood that the improvement contributed to improvement of the usability of Hozos editing operations which would be needed to build high quality ontologies later on. This is why we can evaluate how Hozo was improved in terms of the above items in this experiment.

Table 1 shows the results of the experiment. Due to differences in the number of subjects, all the values of the results were calculated by averaging the values of individual groups. Compared with the version before the improvement, the number of concepts, the number of slots generated, and the number of relational links in the ontology all increased. Furthermore, the number of slot specializations increased considerably. This is conceivably attributed to improvement in slot specialization operations, for which editing was not easy before the improvement. Furthermore, the number of ontology consistency errors decreased, indicating that well-formed definitions of concepts were increased. These results indicate that beginners were able to generate a lot of high quality concepts and relational links. It means that the editing assistance functions contribute to quality assurance for ontology.

Table 1 Results of the examination for evaluating improvement in ontology quality

Subject	number of Basic Concepts	number of Slots	number of Specialized Slots	number of Relational Links	rate of Format Errors
Before improvement (*1)	27.27	21.09	9.82	0.82	10.30%
After improvement (*2)	34.38	29.54	17.54	1.23	4.94%
Rate of change (%)	26.08%	40.05%	78.63%	50.43%	-52.09%

*1: examination before improvement (13 groups, 39 examinees)
*2: examination after improvement (15 groups, 45 examinees)

3 A Method for Content-Based Evaluation

As a quality assurance effort in the refinement phase, it is necessary to evaluate whether the ontology is designed properly in terms of its content (semantics) in addition to the form-based evaluation described in Sect. 2. In this case, methods employed include evaluation by experts in the target domain, verification based on use cases, and evaluation based on the results of application to actual problem solving. A subject of this study is to clarify the method that supports domain experts to confirm and evaluate contents of ontologies. In content-based evaluation, an ontology is evaluated from various viewpoints that include validity of concept definitions and is-a hierarchy classification, validity of relationships among concepts and ability to solve problems. We focus the validity of relationships among concepts because it is one of the main contents of ontology.

For the evaluation of relationships among concepts in an ontology by domain experts, it is important that domain experts can see the relationships from viewpoints according to interests of them. Therefore, Hozo adopts a method for assisting ontology refinement work with a conceptual map generated by exploration and visualization in accordance with the aim of ontology construction and refinement. Although details of the method cannot be described due to space limitation, its outline is as follows. At first, the user selects a concept as starting point for exploring the ontology. Then the system traces relationships (properties) and extracts related concepts according to the user's intention and/or aim. There are several ways to trace relationships. For example, the most primitive one is to follow properties from its domain to range. The user can explore the ontology by choosing combinations of the starting point and the ways to follow relationships. The result of ontology exploration is visualized as a conceptual map. The domain experts can evaluate the ontology thorough the conceptual map instead of checking the ontology directly. This section discusses two cases that we actually applied content-based evaluation with this method to the ontology construction.

3.1 Application to Construction of Sustainability Science Ontology

The knowledge structuring research group at Osaka University Research Institute of Sustainability Science (RISS) has been working on construction of a sustainability science ontology in which knowledge in a variety of domains related to sustainability science in the environmental field is organized in a domain-independent form[3]. In this project, we suggested a basic framework in which maps can be generated from various viewpoints according to interests of experts in an overview of ontology[4].

In the construction phase, the group worked on construction and refinement of the ontology by generating a conceptual map from the constructed ontology and confirming whether the target knowledge was successfully represented as an appropriate map. Furthermore, we applied a similar method in enriching the existing sustainability science ontology mainly with concepts related to the biofuels. 1) First, a domain expert created 29 typical scenarios about production and usage of biofuels before enriching the ontology. 2) Then, on the basis of these scenarios we enriched the existing ontology, to add concepts and relationships appearing in the scenarios. 3) In the refinement tasks, we verified the ontology with the ontology exploration tool, to generate conceptual maps in which the contents of the original scenarios were reproduced (Fig. 1).

As a result, the group was able to perform semantic evaluation of the enriched ontology while maintaining consistency with the existing ontology, demonstrating the effectiveness in the construction and refinement tasks. In these tasks, since the scenarios that are to be represented in the form of a concept map involve a variety of content, there was a strong demand that a method in which the viewpoint used to generate a conceptual map should not have any limitation, and hence our tool has been designed to give users perfect freedom in exploration of the ontology.

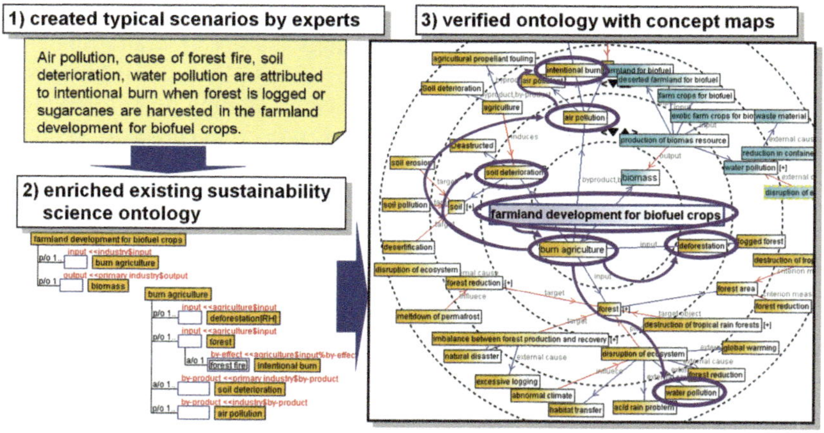

Fig. 1 Example of concept map generated by a typical scenario

3.2 *Concept Map Generation in Construction of Clinical Ontology*

In the Project for Research and Development of Medical Knowledge Infrastructure Database for Medical Information Systems by the Ministry of Health, Labour and Welfare, assuming its indispensable role as an information infrastructure technology for integrated management and sophisticated information analysis of electronic data ranging over multiple domains in clinical medicine, a comprehensive ontology for clinical medicine is being constructed for the first time in Japan [5]. The proposed tool has been extensively used as discussed in the following.

3.2.1 Assistance for Construction and Refinement of Ontology for Connections in Human Body Structure

In the clinical ontology, regarding connective relationships in the human body structure, a common concept construct "Port" is introduced so that various connective relationships, for example, functional connections for passing fluids or signals, such as blood vessels or nerves, mechanical connections, such as connections between bones and joints, and spatial connections representing positional relationships, can be dealt with in isomorphic frames. Accordingly, it is possible to perform reasoning with a computer about links between functional connections, such as the vasculature or the nervous system, or links between bones. In the current implementation, regarding the connective relationships in the human body structure, an ontology for connective relationships involving the circulatory system (arteries and veins), the main nervous systems throughout the body, and the skeletal muscles has been constructed. The numbers of concepts defined in the ontology for the connective relationships are about 11,000 for the circulatory system, including both the arteries and the veins, and about 7,500 for the nervous systems, and the numbers of connective relationships for these systems are about 8,600 and about 3,200, respectively.

In the ontology for connections in the human body structure, in order to confirm whether the connective relationships are properly described in terms of content, the contents described at the destinations of connection ports of the human body constructs are tracked sequentially. As a matter of practice, however, it is substantially impossible to confirm more than 10,000 connective relationships in the human body structure by manually tracking the concept definitions. Thus, in order to confirm that the connective relationships in the human body structure are properly described and to refine the ontology, it is required to perform necessary exploration automatically and to visualize the results in a form easy to recognize for the user.

Thus, in order to meet such a need, we created a special tool that performs exploration to track connection ports representing connective relationships in the ontology describing the human body structure and that visualizes the results for confirmation of the connective relationships in the human body structure (Fig. 2). It means the tool generates conceptual maps by focusing on a common upper level conceptual structure in connective relationships in the human body. This tool is developed as a Java application using JUNG[6]and it supports several visualization forms. In the screen shot shown in Fig. 2, the connective relationship of "Aorta" starting from the "Heart" is visualized. By starting from "Heart" and proceeding to "Vascular connection", it is understood through the visualization that "Aorta" branches at the "Aorta branch" and "Artery" extends to "Femoral artery".

By visualizing the connective relationships of the human body constructs in the clinical ontology as described above, it becomes easier to intuitively understand the description and to find errors in the concept descriptions. Actually, together with experts, we confirmed the connective relationships of "Artery" by using the visualization tool and discovered that a connective relationship of blood vessels that were supposed to be connected had not been described.

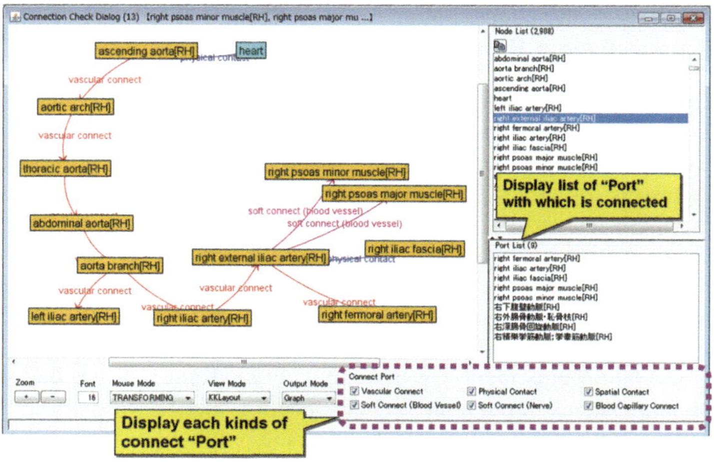

Fig. 2 visualization tool which represent connective relationships in human body structure

3.2.2 Application to Disease Concepts

We also applied the visualization tool described in Sect. 3.2.1 to construction and refinement of disease concepts in a clinical ontology and performed verification. A disease in the clinical ontology is considered as a sum of a series of state transitions, including their causes and intermediate states, and a resulting state caused thereby, and a disease concept is defined as state changes commonly observed among patients who develop the disease. Furthermore, an "Abnormal state" constituting the disease is a generalization of possible states that patients experience, and is defined as a concept having two abnormal states "Cause" and "Result" as attributes. Therefore, it is possible to find all possible abnormal states that patients may experience by tracking "Cause" and "Result" attributes in the abnormal states.

In the current implementation, diseases in twelve diagnostic sections have been defined, amounting to a total of about 6,000 diseases. In refining the disease concepts, tracking and confirming a series of abnormal states that patients may experience was necessary to understand all definitions of a disease concept. In the similar method described in Sect. 3.2.1, a tool for exploring and visualizing state changes among all possible abnormal states in the ontology of disease concepts is created, and it was possible to confirm state changes among the possible abnormal states, as well as the definitions of the disease concepts.

4 Related Work

The method proposed through this research will be further clarified in terms of differences from related research. In form-based evaluation in the ontology construction phase, it is the case with many ontology construction tools that errors are rejected during editing so that an ontology will be edited in compliance with the rules, and consistency is verified after editing. The user must strictly write the ontology without errors during editing, raising a concern that this laboriousness could inhibit conception by the user. Furthermore, OWL-based construction tools often employ reasoners based on description logics (DLs) for consistency verification, such as FaCT++[7], RacerPro[8]and Pellet[9]. In Hozo, consistency verification is performed only minimally during editing, tolerating specific types of errors so as not to inhibit conception by the user, thereby maintaining a certain level of quality of the ontology. As for the items that are to be verified but were not covered during editing, the ontology as a whole is verified after editing. By dividing verification into that during editing and that after editing, the quality of the ontology is assured effectively. Furthermore, since special concept representations are adopted in relation to the reasoner, a special reasoning method is proposed in accordance with the differences in ontology models, and the reasoner is implemented within the construction tool. Hozo Reasoner has 32 native reasoners corresponding to each item for assuring fast consistency-check without logical computation like DLs.

In content-based evaluation in the ontology refinement phase, since judgements based on expert knowledge and experience are required, it is difficult to assist refinement by using a computer. For this matter, in order to assist refinement, proposed approaches for assisting experts understanding of the description of an ontology include an approach in which an ontology is visualized to promote intuitive understanding[10], and an approach in which an ontology is described in a form similar to a natural language by using a control language and the ontology is translated into a formal ontology description through computer processing[11]. There are some approaches that construction tools built into a visualization tool as extension to verify or analyze a ontology such as TGVizTab[12]. However, we focus not on visualization but on exploration of ontologies according to the users' viewpoints. The features of our method are functions for searching for necessary information according to their viewpoint and assisting to understand content of an ontology by expressing conceptual maps for search results. The examples of applications described in this paper demonstrate that the framework itself to search and create conceptual maps at the users' viewpoint is versatile, although partially dependent on the target ontology, and that sufficient effect can be expected. We are sure that a method for content-based evaluation with conceptual maps has sufficient effect.

5 Conclusion

This paper has discussed the authors' efforts with the environment for building/using ontologies: Hozo as a framework that supports quality assurance in the ontology construction and refinement phases, and demonstrated that the methods contributed to quality assurance when actually applied to the ontology construction process.

As a further issue to be addressed, further improvement in the methods for quality assurance in the ontology construction process is needed. Specifically, regarding form-based evaluation, a conceivable approach is to provide appropriate guidance so that even a domain expert who is not familiar with ontology can write properly from the ontology perspective. Regarding content-based evaluation, it is necessary to evaluate our method to be applied to the ontology constraction in other domains. We also have the issue to develop a assistance method for correcting contradictions found in the generated concept map and reflecting the corrections in the ontology.

Acknowledgements. This research was partly supported by the Ministry of Health, Labour and Welfare on Japanese Government as Development business and research of "Medical-knowledge-based database for medical informatics system.", the Environment Research and Technology Development Fund (E-0802) of the Ministry of the Environment, Japan.

References

1. Smith, B., et al.: The OBO Foundry: coordinated evolution of ontologies to support biomedical data integration. Nature Biotechnology 25, 1251–1255 (2007)
2. Hozo, `http://www.hozo.jp`
3. Kumazawa, T., et al.: Toward Knowledge Structuring of Sustainability Science Based on Ontology Engineering. Sustainability Science 4(1), 99–116 (2009)
4. Hirota, T., Kozaki, K., Mizoguchi, R.: Divergent Exploration of an Ontology. In: ISWC 2008, Poster & Demo Notes of the 7th International Semantic Web Conference (2008)
5. Mizoguchi, R., et al.: An Advanced Clinical Ontology. In: Proc. of International Conference on Biomedical Ontology, ICBO, pp. 119–122 (2009)
6. JUNG: Java Universal Network/Graph Framework, `http://jung.sourceforge.net/`
7. FaCT++, `http://owl.man.ac.uk/factplusplus/`
8. RacerPro, `http://www.racer-systems.com/`
9. Pellet, `http://clarkparsia.com/pellet/`
10. Katifori, A., Halatsis, C., Lepouras, G., Vassilakis, C., Giannopoulou, E.: Ontology Visualization Methods–A Survey. ACM Computing Surveys (CSUR), Surv. 39(4), Paper 10 (2007)
11. Dimitrova, V., et al.: Involving Domain Experts in Authoring OWL Ontologies. In: Sheth, A.P., Staab, S., Dean, M., Paolucci, M., Maynard, D., Finin, T., Thirunarayan, K. (eds.) ISWC 2008. LNCS, vol. 5318, pp. 1–16. Springer, Heidelberg (2008)
12. Alani, H.: TGVizTab: An ontology visualization extension for Protégé. In: Proc. of Knowledge Capture (K–Cap 2003), Workshop on Visualization Information in Knowledge Engineering (2003)

Location-Based Web System for Geographically Distributed Mobile Teamwork Management

Eduard-Cristian Popovici, Ioana-Manuela Marcu,
Octavian Fratu, and Simona-Viorica Halunga

Abstract. This paper presents a geographically distributed mobile teamwork management system that combines the simplicity of Web services architecture and protocols, with the GPS-based location information retrieval, and a geographical mapping application. The aim of our work is to offer a Web-based client-server system heading towards a better management of mobile teamwork departments (such as operating and maintenance departments of telecommunications companies) in the idea of making more efficient use of overall resources.

Keywords: location information retrieval, geographically distributed mobile teamwork, GPS, Web system.

1 Introduction

With the widespread availability of cheaper and more powerful portable devices, there is also an increasing demand for services that support seamless communication and collaboration among mobile users. User mobility has also inspired the development of many location-based information services, such as a wide range of new and yet unexplored forms of collaboration, in which information about the user's context, for example, her position, plays a central role in defning both the group of collaborators and the communication mode. The managerial trend of empowering workforces requires mobile workers to rely on a high degree of

Eduard-Cristian Popovici
E.T.T.I. Faculty, University POLITEHNICA of Bucharest,
Iuliu Maniu 1-3, 061071 Bucharest, Romania
e-mail: eduard@elcom.pub.ro

Ioana-Manuela Marcu · Octavian Fratu · Simona-Viorica Halunga
E.T.T.I. Faculty
e-mail: imarcu@radio.pub.ro, ofratu@elcom.pub.ro
shalunga@elcom.pub.ro

E. Mugellini et al. (Eds.): Advances in Intelligent Web Mastering - 3, AISC 86, pp. 217–224.
springerlink.com

teamwork in a changing environment [1]. However, today's mobile workforce managers can't simply pick up an off the shelf solution to realize the teamwork concept for their teams. Current commercial, mobile solutions highlight their main features in terms of data collection, data delivery, and data synchronization between a mobile device and back-end systems. There isn't much mention of effective cooperation among mobile workers, so presumably this isn't regarded as a must-have feature. A few research initiatives could be mentioned in this area. In [2], a middleware architecture (MoCA) with its location inference service (LIS) is described, and an application for context-aware mobile collaboration which is based on this architecture is presented. However, the decentralized approach increases the complexity of the applications. Project MOTION [3], aims to create a flexible, open and scalable information and communication technologies architecture for mobile collaboration. But this environment is concerned to shield from the application developer all aspects regarding mobility and user location, aiming the provision of a seamless, anywhere-available service. In [1] a component-based framework in proposed, for developing agent-based cooperation support systems for mobile workforces, such as personal assistants aren't tied to a particular coordination interaction protocol. The component-based architecture enables personal assistants to dynamically install new cooperative services, with the disadvantage of an increased management complexity. Our work is much a mobile teamwork management oriented system that combines the simplicity of Web services architecture and protocols with the GPS-based location information retrieval and a geographical mapping application, to offer a client-server system heading towards a better management of mobile teamwork departments (such as operating and maintenance departments of telecommunications companies) in the idea of the efficient use of overall resources (Fig. 1).

The paper is organized as follows: Section 2 briefly introduces the architecture of our location-based Web system for geographically distributed mobile teamwork management. Section 3 is concerned with the main components and technologies used to build the mobile teamwork management system. The client application screens for a typical usage of the teamwork tasks management application is described in Section 4. In Section 5 we give some concluding remarks and discuss further plans.

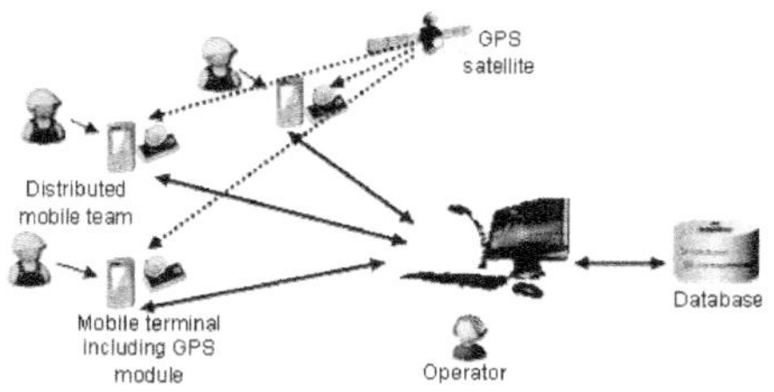

Fig. 1 Location-based Web system for distributed mobile teamwork management

2 The High-Level Architecture of the System

The geographically distributed mobile teamwork management based on location services on mobile phone is intended for departments that have activities involving travel from one area to another. We take the example of a company that has these teams grouped by geographical areas in the country. Teams are assigned different tasks at a time by their superior, the head office, according to the geographical area where they belong. The simplified architecture of our system can be divided into three different tiers:

- the phone (client) that uses the mobile application (user interface) and has a GPS module, either integrated or external;
- the Web server (hosting the user interface and the web services);
- the database server that hosts all the relevant information (Fig. 2).

Application to field teams runs on a mobile phone and server-side software will be installed on a computer (including database and application servers). The application architecture and the server synchronization mechanism are implemented considering the portability. For this reason the operator application has a browser-based web interface. User interfaces are intuitive, easy to use and consistent for client and server.

3 The Main Components and Technologies

The main technologies used to build the distributed mobile teamwork management system are open source ones. In the following we present these technologies starting from client and ending with database.

3.1 Client Components and Technologies

For the client application, called MobileWorkAssistant, we used Java Mobile Edition configured for any phone with CLDC 1.1 and having a MIDP 2.0 profile [4].

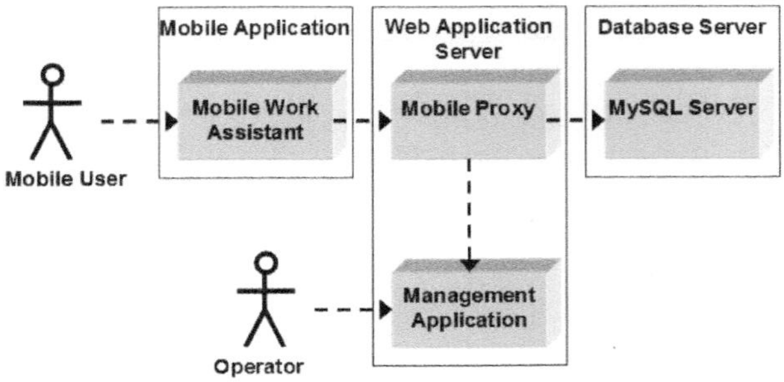

Fig. 2 The simplified architecture of our location-based Web system for geographically distributed mobile teamwork management

In this way the application has the advantage that it can be installed and used on almost any phone with Java Virtual Machine. In addition, it represents a very inexpensive solution to companies (almost no additional investment in equipment). The application has the following sequence of execution:

1. Download the original settings connection to the server.
2. Login (for simplicity and safety)
3. Search and establish a communications with GPS receiver (NMEA protocol)
4. Collecting data from the GPS device and obtaining geographic coordinates by choosing and parsing $ GPGGA and $ GPRMC sentences
5. Sending the coordinates to the server. (Continuing on a separate thread)
6. Posting of personal data and information about the last use.
7. Synchronization tasks for team field
8. Viewing information for each started task:
    - destinations, regions, locations of deposits of materials, etc.
    - work order description, needed equipments, etc.
9. Trace and execute the tasks

Packages of MobileWorkAssistant client application and communication between them are described in the (Fig. 3), the arrows indicating calls. The application instantiates the controller class and starts it on a separate thread. Then it searches GPS devices by using GPSDevice package and Bluetooth classes for communication. After discovering and establishing a connection with the device, it starts the thread execution for acquisition of data sent by the GPS receiver. This information is received in the form of sentences with the format described in NMEA protocol [5].

Propositions needed to retrieve latitude and longitude in the format required invoking the Virtual Earth Web service [6] on server to insert pushpins on a map, are $ GPGGA and $ GPRMC. These are parsed in the GpsPositionParser.java class and the coordinates are obtained. The access to Web services is done by means of the

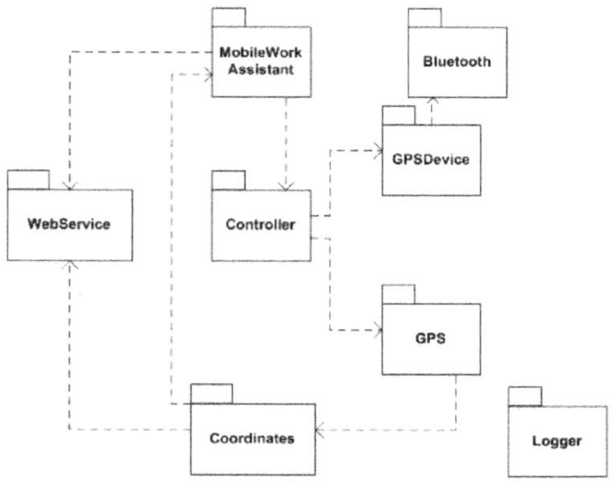

Fig. 3 The packages of the MobileWorkAssistant client application

the Java ME Web Services API [7]. Developed within the Java Community Process as JSR 172, the Java ME Web Services API (WSA) extends the Java Micro Edition to support web services. WSA is designed to work with Java ME profiles based on either the Connected Device Configuration (CDC) or the Connected Limited Device Configuration (CLDC 1.0 or CLDC 1.1). The goal of WSA is to integrate fundamental support for web services invocation and XML parsing into the device's runtime environment, so developers don't have to embed such functionality in each application - an especially expensive proposition in resource-constrained devices like mobile phones and personal digital assistants. JSR 172 specifies standardized client-side technology to enable Java ME applications to consume remote services on typical web services architectures, as Fig. 4 illustrates.

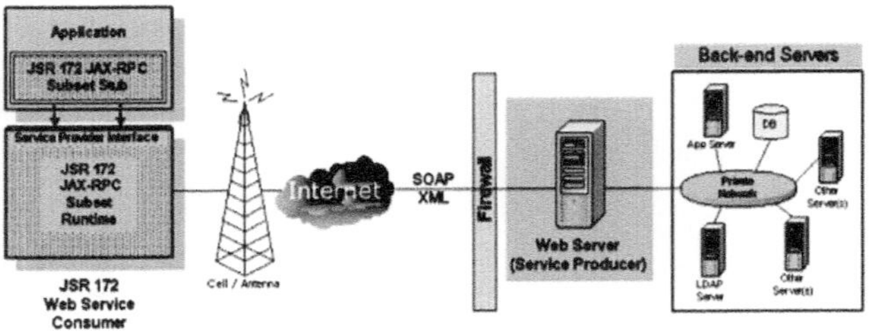

Fig. 4 JSR 172 specifies standardized client-side technology to enable Java ME applications to consume remote services on typical web services architectures

3.2 Server Components and Technologies

Server technologies used are Java Server Faces (JSF) [8] and JavaServer Pages (JSP) [9]. The application is installed on a Java Web server that provides services necessary for both desktop and mobile applications. The desktop activity management application, WorkManager, to allow access to it from any system connected to the Internet. For this, it has a Web interface, implemented with JSP and JSF Web technologies. A sample of this interface for work orders planning is shown in Fig. 5.

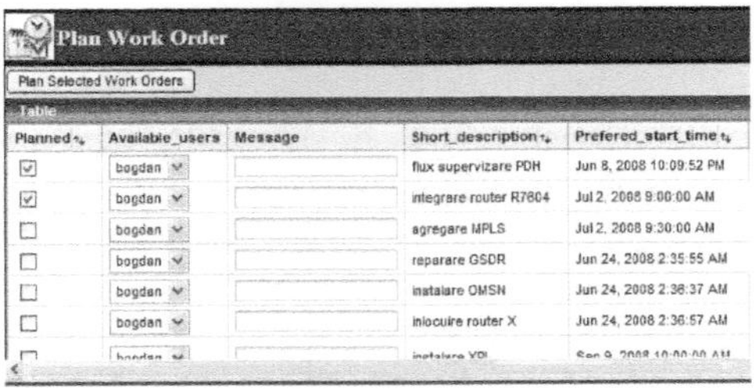

Fig. 5 The Web interface for work orders planning

The application includes a mapping solution for GPS position of the field teams on a map, for viewing or watching them, called the Virtual Earth SDK from Microsoft [6], that provides free services - MapPoint Web Services - to display various information (picture, name, time, etc.) on a map. Fig. 6 shows the Virtual Earth map, decorated with information obtained from the client's GPS location service.

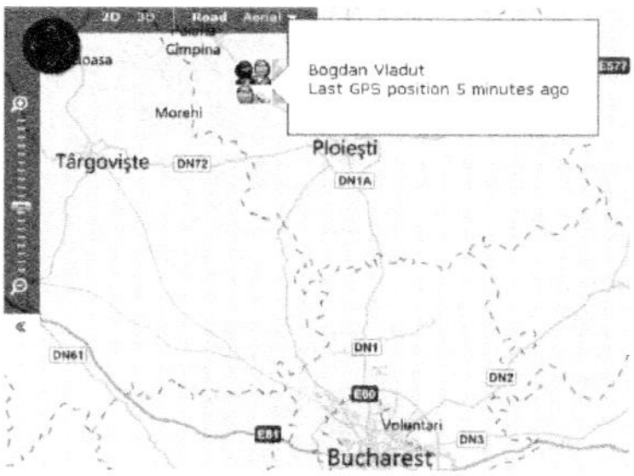

Fig. 6 The map offered by Virtual Earth used to show the position of a team member

4 Typical Usage of the Teamwork Tasks Management Application

When starting the MobileWorkAssistant mobile application, it starts looking for GPS device. After successfully connecting to the GPS module, the screen will display the geographical coordinates received from GPS module, at the time. After logging, the mobile application extracts from the database the recorded profile data, i.e. name, first name and last visit (date of last login and logout), as shown in Fig. 7.

Fig. 7 The team member profile shown on the mobile phone screen

In the same way, will be extracted the tasks the team member has to fulfill (every task is defined by a brief description, start time of task execution, and the type of task) as shown in Fig. 8. When the button "Do task" is pressed, the first task that the operator associated to him will be detailed, in order to be executed. Location details are displayed, such as a broader description, equipment needed, and additional information.

The overall system was field tested, with mobile phones connected through Bluetooth connections to external GPS modules, and equipped with Java Mobile Edition, connected to a Java Web Server called Glassfish 2.0 that generates the JSP and JSF Web interfaces, and provides access to the Web services.

Fig. 8 The tasks list associated to a team member, and a sample of the task details, shown on mobile phone screens

5 Conclusion

In this paper, we presented a geographically distributed mobile teamwork management system that combines the simplicity of Web services architecture and protocols, with the GPS-based location information retrieval, and a geographical mapping application. The aim of our work was to offer a Web-based client-server system heading towards a better management of mobile teamwork departments (such as operating and maintenance departments of telecommunications companies) in the idea of making more efficient use of overall resources. The field tests realized with mobile phones connected through Bluetooth connections to external GPS modules proved to be easy to use, intuitive and portable, on a broad range of devices. Though, a better experience is expected to be obtained using smartphones with integrated GPS support. We plan to extend the mobile application by incorporating new services such as a navigation guiding service, for team members to easy get to the location task. In the management (back office) application, a very useful service would be the generation of periodical (monthly, yearly, etc.) work related reports, statistics on various criteria that could be used to improve the activities. We also started to explore some technological alternatives to Java Mobile Edition mobile platform, such as the Android mobile platform [10] ; and some technological alternatives to the classical Web services (SOAP-based) [7] , such as the RESTful Web services [11] .

Acknowledgements. This work is sponsored by Romanian Authority of Scientific Research through the 81-026/2007 "New Services and Applications for satellite navigation and location using WiMAX technology (LOCOMAX)" project and the 12-126/2008 "Hybrid wireless access system with unique addressing (SAWHAU)" project.

References

1. Lee, H., Mihailescu, P., Shepherdson, J.: Realizing Teamwork in the Field: An Agent-Based Approach. IEEE Pervasive Computing 6(2), 85–92 (2007)
2. Rubinsztejn, H.K., Endler, M., Sacramento, V., Gonçalves, K., Nascimento, F.: Support for Context-Aware Collaboration. In: Karmouch, A., Korba, L., Madeira, E.R.M. (eds.) MATA 2004. LNCS, vol. 3284, pp. 37–47. Springer, Heidelberg (2004)
3. Dustdar, S., Fenkam, P.: Formally designing Web services for mobile team collaboration. In: Proceedings of 30th Euromicro Conference, Rennes, France, August 31-September 3, pp. 469–476 (2004)

4. Java ME Technical Documentation,
http://download.oracle.com/javame/index.html
5. GPS Navstar Joint Program Office, NAVSTAR GPS User Equipment Introduction
http://www.navcen.uscg.gov/pubs/gps/gpsuser/gpsuser.pdf
(1996)
6. Microsoft Virtual Earth API,
http://www.programmableweb.com/api/microsoft-virtual-earth
7. Introduction to J2ME Web Services,
http://developers.sun.com/mobility/apis/articles/wsa/
8. JavaServer Faces Technology - Documentation,
http://www.oracle.com/technetwork/java/javaee/documentation/index-137726.html
9. Servlets and JSP Documentation,
http://www.oracle.com/technology/docs/tech/java/servlets
10. Android Platform Developer's Guide,
http://developer.android.com/guide/index.html
11. RESTful Web Services,
http://en.wikipedia.org/wiki/Representational_State_Transfer

Two New Methods for Network Analysis: Ant Colony Optimization and Reduction by Forgetting

Václav Snášel, Pavel Krömer, Jan Platoš, Miloš Kudělka,
Zdeněk Horák, and Katarzyna Wegrzyn-Wolska

Abstract. This paper presents two new methods for network analysis. Ant colony optimization is a nature inspired algorithm succesfull in graph traversal and network path finding whereas network reduction based on stability introduces two new properties of network vertices based on their long-term behavior, their role in the network and the understanding of how memory works. We illustrate the algorithms on applications in social network analysis and information retrieval using the DBLP dataset and a small network of hyperlinked documents.

Keywords: Network analysis, memory, stability, complexity reduction, ant colony optimization, search personalization.

1 Introduction

Network analysis becomes hot topic as many real world entities are studied from the perspective of their relations and interaction. The dynamics in time, evolution, and growth of networks are investigated. In this work, we investigate a network of hyperlinked documents and use ant colony optimization to personalize hyperlink matrix for relevance estimate propagation. Next, we study ability of forgetting curve, retention, and stability to reduce a co-authorship network and find authors with strong ties.

Václav Snášel · Pavel Krömer · Jan Platoš · Miloš Kudělka · Zdeněk Horák
VŠB - Technical University of Ostrava,
17. listopadu 15, 708 33, Ostrava-Poruba, Czech Republic
e-mail: {vaclav.snasel,pavel.kromer,jan.platos,milos.kudelka,
zdenek.horak}@vsb.cz

Katarzyna Wegrzyn-Wolska
ESIGETEL, Avon, France
e-mail: katarzyna.wegrzyn@esigetel.fr

E. Mugellini et al. (Eds.): Advances in Intelligent Web Mastering - 3, AISC 86, pp. 225–234.
springerlink.com

2 Ant Colony Optimization

Ant colony optimization (ACO) is a popular meta-heuristic algorithm based on certain behavioral patterns of foraging ants. Emulation of ants' behavior can be used as probabilistic computational technique for solving complex problems which can be reduced to finding optimal paths [1, 2]. An artificial ant k placed in vertex i moves to node j with probability p_{ij}^k:

$$p_{ij}^k = \frac{\tau_{ij}^{\alpha}\eta_{ij}^{\beta}}{\sum_{l \in N_i^k}(\tau_{il}^{\alpha}\eta_{il}^{\beta})} \tag{1}$$

where N_i^k represents the neighborhood of ant k in node i (i.e. nodes that are available to move on), τ_{ij} represents amount of pheromones placed on arc a_{ij} and η_{ij} corresponds to a-priori information reflecting the cost of passing arc a_{ij}. After the ants finish their movement forward, they return to the nest with food. The tour of $k-$th ant is denoted as T^k. The length of T^k called C^k or the amount of collected food L_k (i.e. solution quality) is used to specify the amount of pheromones to be placed by ant k on each arc on the trail that led to the food source:

$$\tau_{ij} = \tau_{ij} + \sum_{k=1}^{m} \Delta\tau_{ij}^k, \qquad \Delta\tau_{ij}^k = \begin{cases} \frac{1}{C^k} & \text{if arc } (i,j) \text{ belongs to } T^k \\ 0 & \text{otherwise} \end{cases} \tag{2}$$

After all ants finish one round of their movement, the pheromones evaporate (i.e. the amount of pheromones on each arc is reduced):

$$\tau_{ij} = (1-\rho)\tau_{ij} \tag{3}$$

The coefficients α, β and ρ are general parameters of the algorithm. This initial version of ACO algorithm is called ant system (AS).

2.1 Implicit Relevance Based Inquirer Model

In this work, ACO is used to personalize the weights of hyperlinks in a network of connected documents. Subsequently, the hyperlink weights are used to propagate relevance estimate from one document to another.

In the implicit inquirer model, we estimate the relevance of documents by the number of times the user clicked them, by click-through data. However, the documents in a hyperlinked environment are not standalone. They link other documents and they are linked from other documents. A user who displays document d after she or he clicked it might be interested also in documents that are linked from d or that link to d. A click increases the relevance estimate $r(u,d)$ but it might increase the relevance estimate of linked documents too. When the user follows some of the hyperlinks leading from d, she or he can click those potentially relevant. In that case, their relevance estimate should be increased more than the relevance estimate for the non-clicked hyperlinks. In order to take advantage of the implicit information encoded into the network of a hyperlinked collection, we propagate the relevance estimate over hyperlinks.

Imagine a collection of hyperlinked documents as a multigraph $G = (V, E)$ where the set of vertices V corresponds to the set of documents D and the set of edges E corresponds to document hyperlinks. In (4) we define a general $k-$step document relevance estimate vector for user u.

$$r_k^T = r^T + \sum_{l=1}^{k} r^T H_u^l \tag{4}$$

where r is a click-through relevance estimate vector containing user specific relevance estimates for all documents in D and H_u is user specific hyperlink matrix. The value of $H_{ij} \in [0,1]$ represents the factor affecting relevance propagation from node i to node j. The higher H_{ij} the stronger the relevance propagation from i to j. For $k = 1$, the relation from (4) takes the form

$$r_1^T = r^T + r^T H_u \tag{5}$$

and represents a situation in which relevance estimate of every document d consists of $r(u,d)$ and relevance propagated from nodes that immediately link to d. For $k = 2$, the relevance estimate would be also affected by relevance propagated from nodes whose path to d has the length 2 and so on.

3 Forgetting Curve

Generally, the term memory is understood as committing, storing and recalling of experiences. The role of memory is crucial because it stores and recalls all the information we need for our normal lives. All stimuli and situations in which we find ourselves are compared to their traces in memory, which allows us to recognize the meaning of these stimuli and situations. Recalling information is either a reproduction or re-memorization of already known information. The process of forgetting is opposite to the process of recalling. To forget something means not to lose the particular memory trace, but replace it with a new experience. Nothing is forgotten, it just cannot be recalled, because it has lost its meaning. There are two factors causing information to be forgotten. The first one is the extinction of the unused memory trace and the second is the interference of new experiences - the replacement of less important information by the more important ones.

Ebbinghaus proposed the forgetting curve in 1885. The forgetting curve (see [5]) defines the probability that a person can recall information at time t since previous recall. It can describe long-term memory and is usually presented using the following equation.

$$R = e^{-\frac{t}{S}}$$

R (memory retention) the probability of recalling information at time t since the last recall.

t time since the last recall.

S (relative strength of memory - stability) approximated time since the last recall for which is the information stored in memory.

Remark. There are also different approaches for the computation of the forgetting curve (see for example [6]) but the conclusions are always very similar - the forgetting process is much faster in the beginning (see fig. 1).

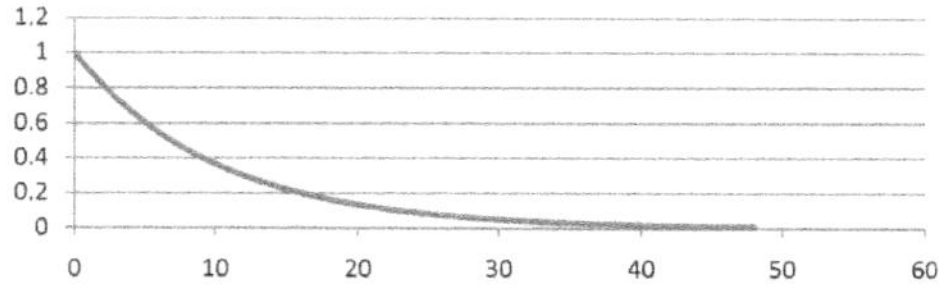

Fig. 1 Forgetting Curve

The computation depends on the type of memory, especially on the estimated time S (this value is not constant in the long term). For simplicity, assume that if we work with the information for the first time, then the time of storing information in memory is $S_{ini} > 0$ and this default value is constant.

An important feature of long-term memory is that after reproduced information recall in the time $t > 0$, the time of storing information in memory S changes. The change is dependent on the previous time S and on the time of recall t. Ideally, the reproduced recall multiplies this time (in comparison with the previous value) by factor $F > 1$.

The other important feature of long-term memory is, that immediate reproduced recall (too quick) of information has no bigger effect on the learning. On the other hand, the reproduced recall too lately (in time near S) causes substantial forgetting. There is an optimal time between these two extreme situations in which the reproduced information recall causes a high level of remembering (and consequently the maximum increase of time S by factor F).

In the ideal case (reproducing the information in optimal time), the remembering of information is gradual and very effective - after each recall, the time of storing information in memory S (remembering) is multiplied by factor F. For updated S_{new}, after new information recall should hold:

1. If $t > S$ then $S_{new} = S_{ini}$ (information is considered as new)
2. If $t \to S$ then $S_{new} \to S_{ini}$ (late recall is considered as almost new information).
3. If $t \to 0$ then $S_{new} \to S$ (early recall has almost no influence)
4. If $t \to \text{opt}(S)$ then $S_{new} \to F \cdot S$, where $\text{opt}(S)$ is the function returning optimal time for recalling the information and F is the factor of optimal improvement.

Remark. For reproduced information recall is $R = 1$. This follows from the fact, that $t = 0$ at this moment. For the factor of optimal improvement holds, that when the information is recalled at optimal time, the value of S is multiplied by two (depending on the type of memory). Therefore we can assume that $F \in (1;2\rangle$.

We have to consider three things:

1. The function $\mathrm{opt}(S)$ for the calculation of optimal information recall time.
2. The choice of optimal improvement factor F.
3. Function $f(t,S,F)$ for calculation of S_{new}.

Function opt(S). Available sources present the optimal time for reproduced information recall in the range of 10–30% of time S. The setting of this function is dependent on the type of memory (e.g. $\mathrm{opt}(S) = 0.2 \cdot S$).

The factor F of optimal improvement. The factor F is involved in the computation of time S for which the information is held in memory (is remembered). This factor is again dependent on the type of memory. For the calculation of S with the same type of memory the value of F is constant (e.g. $F = 1.2$).

The function $ch(t,S)$. The value of S_{new} is dependent on the type of memory, on the time of repetitive information recall and on the previous value of S (this incorporates the history of learning mentioned information). For the calculation of S_{new} we need to design the function of $ch(t,S,F,S_{ini})$ for the calculation of the coefficient of change of the value S. Then holds:

$$S_{new} = ch(t,S,F,S_{ini}) \cdot S$$

Available sources contains various approaches for the computation of value of function $ch(t,S,F,S_{ini})$. For example we will use simple relation based on linear functions (see fig. 2):

1. If $0 \leq t \leq \mathrm{opt}(S)$ then $ch(t,S,F,S_{ini}) = 1 + (F-1) \cdot \frac{t}{\mathrm{opt}(S)}$
2. If $\mathrm{opt}(S) \leq t \leq S$ then $ch(t,S,F,S_{ini}) = F - (F - \frac{S_{ini}}{S}) \cdot \frac{t-\mathrm{opt}(S)}{S-\mathrm{opt}(S)}$
3. If $t > S$ then $ch(t,S,F,S_{ini}) = \frac{S_{ini}}{S}$

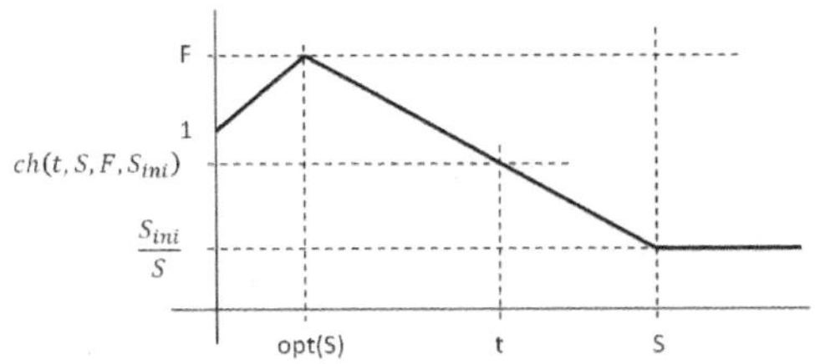

Fig. 2 Calculation of S in time t

3.1 Forgetting of Social Network

We assume that interactions between particular pairs of vertices take place in the social network continuously. If we consider these interactions as an experience stored in memory, then the ties between two vertices of the network are more stable, if this network learns these interactions. As a result we assume that the more interactions occur between the two vertices, the more stable is the tie between them. Therefore we can understand the social network as a set of variously stable ties.

Remark. Interaction between two vertices as well as information leaves traces in memory. This trace is dependent on how often these interactions take place (as an analogy to the reproduced information recall). If we will understand the network as an analogy to the human brain, then the memorization of ties in the network will correspond to the degree of remembering the information in the brain.

The properties of ties change over time, depending on how often and in what time two vertices interact. For the calculation of the properties of ties we use the Forgetting Curve. It is the analogy to the learning and forgetting of reproduced information - reproduced interaction. For each tie we define three time-changing characteristics.

Definition 1 (Edge Retention, Edge Stability, Active Edge). ***Edge Retention*** ER *expresses the probability that a reproduced interaction will take place in given time* t *between two vertices connected by given edge.* ***Edge Stability*** ES *is the estimated time for which the tie between vertices remains active (since given time* t*).* ***Active Edge*** *is a tie, for which holds that* $ES > 0$ *in given time* t*.*

Like the retention and stability of ties we can define the same vertex characteristics and use the forgetting curve in their calculations again.

Definition 2 (Vertex Retention, Vertex Stability, Active Vertex). ***Vertex Retention*** VR *expresses the probability that a reproduced interaction will take place in given time* t *between this vertex and any other vertex.* ***Vertex Stability*** VS *is the estimated time for which the vertex remains active (since given time* t*).* ***Active Vertex*** *is a vertex, for which holds that* $VS > 0$ *in given time* t*.*

4 Experiments

We illustrate the presented methods on experiments with some real-world networks.

For the ACO experiments, we have used a dump of one language mutation of the popular Web encyclopedia Wikipedia[1]. Simple English Wikipedia[2] is a language mutation of the Wikipedia written in simple English with limited vocabulary and using only simple grammar. For our purposes, it is attractive because it contains only a relativelly small number of articles in English language. As in other language variants, articles are linked (within simple Wikipedia but also to other Web sites) and categorized.

[1] http://wikipedia.org/

[2] http://simple.wikipedia.org/

We have selected small subset of Simplewiki to perform experiments in order to tune the genetic programming for evolutionary query optimization. We have used random node sampling method to select 110 documents (nodes) from Simplewiki at random with uniform probability. Selected nodes and links between them formed a small test collection called Rand110. An information retrieval system over Rand110 has been implemented using the extended Boolean information retrieval model [3]. Gerard Saltons $tf \times idf_t$ indexing mechanism [4] has been used to create vectors of weighted index terms representing documents in Rand110.

We have investigated effects of relevance propagation by (4) in test collection Rand110. We have taken the query "month or year or day" and used it to mark the fuzzy set of relevant documents. Fuzzy set of relevant documents serves as an initial relevance estimate r. Fig. 3 shows how the relevance estimate for some documents increases with growing k.

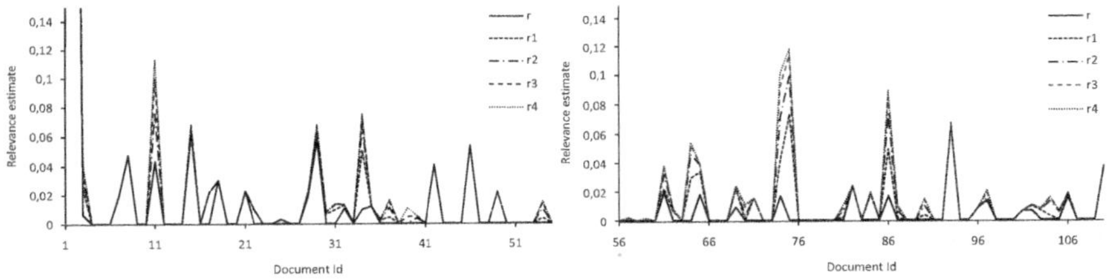

(**a**) Relevance propagation for the first 55 documents in Rand110.

(**b**) Relevance propagation for the last 55 documents in Rand110.

Fig. 3 Example of $k-$ step relevance propagation in Rand110 for $k \in \{1,2,3,4\}$

The relevance estimate consists of the sum of propagated click-through relevances of documents that are connected to d by a link or path and the click-through score of d itself. An important part of (4) is the personalized hyperlink matrix. In contrast to methods based on linear algebra such as PageRank or HITS, we use ant colony optimization to create and maintain user specific document hyperlink matrix. Ant colony optimization is a suitable tool for discovery and maintenance of user specific relationships among documents. It provides means for discovery of novel paths through ants' foraging as well as supression of no longer used links through pheromone evaporation. In general, an update to H_u can be denoted as

$$H_{(i+1)u} = aco_H(H_{(i)u}, r_k) \tag{6}$$

The ant colony optimization algorithm aco_H updates the personalized hyperlink matrix according to its current state and current ($k-$step) document relevance estimate. Aco_H places ants on graph vertices randomly with probability proportional to the relevance estimate of corresponding documents. We let the ants move n steps forward in the graph. We set the number of steps the ants should move equal to the value of graph diameter, but it can be customized. After their run, the ants are rewarded according to the sum of relevance estimate of documents they have visited on their trail and the pheromones describing probability of transition from node i to node j (i.e. the values of H_{ij}) are updated. The operation is repeated with all ants N times. The ants are seeking for paths between most relevant documents and they discover how are the documents organized. By assigning higher weight (higher amount of pheromones) to hyperlinks on paths between most relevant documents, the ants suggest directions in which the relevance estimate should be propagated.

After a period of user activities, the aco_H algorithm can be used to update users hyperlink matrix. An example of the evolution of link weights in Rand110 is shown in Fig. 4. Document links are displayed as a rectangular matrix with dimension 110×110. The color of pixel at the position (i, j) represents the weight of link between document i and document j. The darker the color, the stronger the connection between i and j. Yet again, let us remind that the updates of personalized hyperlink matrix are subject to ACOs stochastic nature. It means that $aco_H(H_{(i)u}, r_k)$ might result in slightly different $H_{(i+1)u}$ in independent runs.

The number of ants, trial length and values of α and ρ affect the way aco_H updates the personalized hyperlink matrix, but this is also subject to concrete application requirements (e.g. how long paths shall be explored? how fast should the hyperlink weights evaporate?). In our examples, α was set to 1 because no a-priori information η_{ij} was used. Sample $aco_H(H_{(i)u}, r_k)$ outputs for different values of ρ are shown in Fig. 4. The query "month or year or day" was used to mark relevant documents and r_4 relevance estimate was used as r_k. We can clearly see how higher evaporation rate results in more sparse hyperlink matrices.

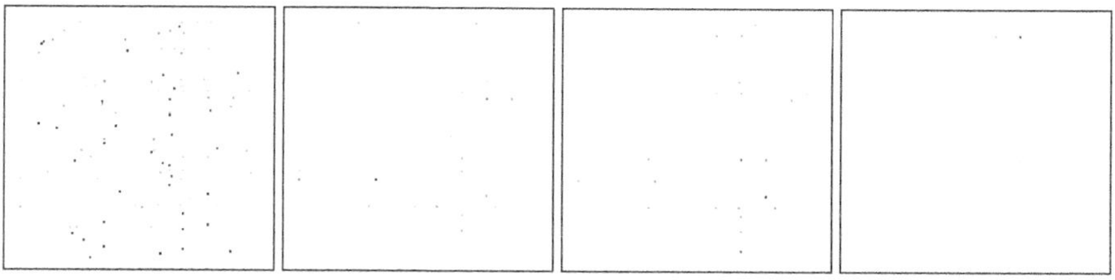

(a) Rand110 row normalized hyperlink matrix $H_{(i)u}$. **(b)** $aco_H(H_{(i)u}, r_k)$, $\rho = 0.01$. **(c)** $aco_H(H_{(i)u}, r_k)$, $\rho = 0.1$. **(d)** $aco_H(H_{(i)u}, r_k)$, $\rho = 0.5$.

Fig. 4 Example of Rand110 link weights adjustment by aco_H

For the experiments with forgetting curve, we need time-dependent data to calculate the retention and stability. In order to obtain such a data, we downloaded the DBLP dataset from April 2010 in XML[3] and preprocessed it for further usage. First

[3] Available from `http://dblp.uni-trier.de/xml/`

of all, we selected all conferences held by IEEE, ACM or Springer, which gave us 9,768 conferences. For every conference we identified the month and year of the conference.

In the next step we extracted all authors having at least one published paper in the mentioned conferences (as authors or co-authors). This gave us 443,838 authors. Using the information about authors and their papers we were able to create a set of cooperations between these authors consisting of 2,054,403 items. An important fact is that *cooperation* is understood to be the co-authorship of one paper. Using the information about the conference date, we accompanied these cooperations by time information. We also ignored the ordering of author names as it is impossible to investigate the particular ordering protocol (by alphabet, by contribution, etc.) and hence all co-authors are given equal importance.

We computed the weight of edges and vertices as their stability in time t. We divided the entire recorded publication period of conferences (the first record from 1963) into one-month time periods. If during one month an author has published a paper with another co-author in at least one conference (held by IEEE, ACM or Springer), then we set one interaction for the both authors (vertices) and the tie between author and co-author (edge) for this month. For each vertex and edge we obtain a list of months in which the interactions occurred. Then we applied the forgetting curve to compute the retention and stability of every author and tie in a specified month.

We have truncated the selected time period to December 2008 to obtain the most complete dataset. Of course, the weight (stability of vertices and edges) changes in every month, but the following calculations are made until the end of year 2008. At that time only 122 289 authors and 248 519 ties were active (the other had a stability equal to zero). Stability does not depend only on the number of interactions (and the number of publications consequently) but also depends on how often and how regularly these interactions occur. The calculation of retention and stability for each vertex and edge has linear time complexity with the number of interactions. Consequently, the calculation is very effective even for large networks.

As a result of our experiment, in fig. 5 can be seen selected authors with strong ties to co-authors and their co-authors. Weak ties was filtered and most stable relations describe a regular cooperation between authors.

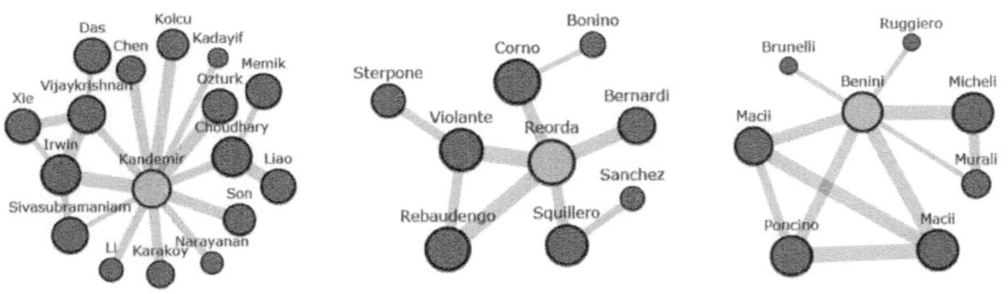

Fig. 5 Selected authors with strong ties to surroundings

5 Conclusions

Ant colony optimization and forgetting curve are promising methods for the analysis of complex networks.

In information retrieval, the weighted links between documents are used to propagate the relevance estimate through the set of all documents. To reflect users individual view on the hyperlinked structure and to consider individual importance of document links, ant colony optimization has been used for inference and maintenance of user specific hyperlink matrix and discovery of novell paths for relevance propagation among documents in the collection.

For the social network reduction, two new vertex and edge parameters called the retention and the stability were introduced. They were calculated with the help of the forgetting curve, which is a well-known approach as a result of experiments with human memory. It is used as a heuristic, which allows us to effectively analyze the stability of elements of a social network and also to reduce the network to the most important components. The network works then in a similar way as the human brain, which forgets information and also learns new things.

Acknowledgement. This work was supported by the Czech Science Foundation, under the grant no. GA201/09/0990.

References

1. Dorigo, M., Stützle, T.: Ant Colony Optimization. MIT Press, Cambridge (2004)
2. Abraham, A., Guo, H., Liu, H.: Swarm intelligence: Foundations, perspectives and applications. In: Nedjah, N., de Macedo Mourelle, L. (eds.) Swarm Intelligent Systems. SCI, vol. 26, pp. 3–25. Springer, Heidelberg (2006)
3. Crestani, F., Pasi, G.: Soft information retrieval: Applications of fuzzy set theory and neural networks. In: Kasabov, N., Kozma, R. (eds.) Neuro-Fuzzy Techniques for Intelligent Information Systems, pp. 287–315. Springer, Heidelberg (1999)
4. Salton, G., Buckley, C.: Term-weighting approaches in automatic text retrieval. Information Processing and Management 24(5), 513–523 (1988)
5. Ebbinghaus, H., Ruger, H.A., Bussenius, C.E.: Memory: A contribution to experimental psychology (1885/1913)
6. Wixted, J.T., Ebbesen, E.B.: Genuine power curves in forgetting: A quantitative analysis of individual subject forgetting functions. Memory and Cognition 25, 731–739 (1997)

Author Index

GPSR Compliance

The European Union's (EU) General Product Safety Regulation (GPSR) is a set of rules that requires consumer products to be safe and our obligations to ensure this.

If you have any concerns about our products, you can contact us on ProductSafety@springernature.com

In case Publisher is established outside the EU, the EU authorized representative is:

Springer Nature Customer Service Center GmbH
Europaplatz 3
69115 Heidelberg, Germany

Batch number: 10755090

Printed by Printforce, the Netherlands